Muslim Voices a
Contempor

Previous Publications

Frances Trix

Spiritual Discourse: Learning with an Islamic Master (1993)
Albanians in Michigan: A Proud People of Southeast Europe (2001)
"Kosovar Albanians between a Rock and a Hard Place," in Ramet and Pavlakovic (eds.). *Serbia Since 1989: Politics and Society Under Milosevic And After* (2005)

John Walbridge

The Science of Mystic Lights: Qutb al-Din Shirazi and the Illuminationist Tradition in Islamic Philosophy (1992)
Sacred Acts, Sacred Time, Sacred Space: Essays on Bahá'í Law and History (1996)
The Philosophy of Illumination of Shihab al-Din Yahya Suhrawardi (with Hossein Ziai) (1999)
The Leaven of the Ancients: Suhrawardi and the Heritage of the Greeks (2000)
The Wisdom of the Mystic East: Suhrawardi and Platonic Orientalism (2001)

Linda Walbridge

Without Forgetting the Imam: Lebanese Shi'ism in an American Community (1997)
The Most Learned of the Shi'a: The Institution of the Marja' Taqlid (2001)
The Christians of Pakistan: The Passion of Bishop John Joseph (2002)
Personal Encounters: A Reader in Cultural Anthropology (with April K. Sievert) (2002)

Muslim Voices and Lives in the Contemporary World

Edited by

Frances Trix, John Walbridge,
and
Linda Walbridge

palgrave
macmillan

MUSLIM VOICES AND LIVES IN THE CONTEMPORARY WORLD

Softcover reprint of the hardcover 1st edition 2006 978-0-230-60536-7

Photo by John Walbridge Iranian woman restoring Safavid mural, Isfahan, Iran, 2006

First published in 2008 by
PALGRAVE MACMILLAN™
175 Fifth Avenue, New York, N.Y. 10010 and
Houndmills, Basingstoke, Hampshire, England RG21 6XS
Companies and representatives throughout the world.

PALGRAVE MACMILLAN is the global academic imprint of the Palgrave Macmillan division of St. Martin's Press, LLC and of Palgrave Macmillan Ltd. Macmillan® is a registered trademark in the United States, United Kingdom and other countries. Palgrave is a registered trademark in the European Union and other countries.

ISBN 978-1-349-37282-9 ISBN 978-0-230-61192-4 (eBook)

DOI 10.1057/9780230611924

Library of Congress Cataloging-in-Publication Data

Muslim voices and lives in the contemporary world / edited by Frances Trix, John Walbridge, Linda Walbridge.
p. cm.
Includes bibliographical references and index.
ISBN 978-0-230-60536-2
1. Muslims—Social conditions—21st century. I. Trix, Frances. II. Walbridge, John. III. Walbridge, Linda S.

BP163.M9363 2008
305.6'97—dc22 2007035890

A catalogue record for this book is available from the British Library.

This book is printed on paper suitable for recycling and made from fully managed and sustained forest sources. Logging, pulping and manufacturing processes are expected to conform to the environmental regulations of the country of origin.

Design by Newgen Imaging Systems (P) Ltd., Chennai, India.

First edition: May 2008

10 9 8 7 6 5 4 3 2 1

For Linda Strickland Walbridge (1946–2002)
who studied Muslims
newly at home in America

Contents

List of Figures

Preface

We had the good fortune to live in Muslim communities in quieter times. We got to know Muslims of our own age, their parents, and grandparents. We lived in capital cities, villages, and immigrant communities. We knew Muslim students, teachers, clerics, shopkeepers, landlords, professionals, workers, neighbors, and their families. We also studied their languages—Arabic, Turkish, Persian, Urdu, and Albanian in our cases—and so experienced support and understanding, and sometimes humorous misunderstanding, as we worked at interacting through sounds and grammar and phrases that were very different from those of our Midwestern American upbringing. We found ourselves returning again and again and count the people we met, with whom we have maintained contact, among our friends. But since the 1990s, and especially since 9/11, we have found our experience increasingly at odds with the way Muslims are portrayed in the American media.

In this book we present personal encounters with individual Muslims. These Muslims are neither famous nor infamous; rather, they are people from different fields and various walks of life—from a religious musician to a pediatrician, from a village healer to a visual artist, from a religious leader to a coordinator of a local relief agency. They belong to both Sunni and Shi'a Muslim communities, and live in different countries in Europe, Africa, the Middle East, South Asia, Southeast Asia, and North America. They include Muslims grounded in traditional ways of life, Muslims who have been displaced and who have had to adjust to life in new places or circumstances, and Muslims who have reconciled their lives to changing circumstances in new ways.

Each of the eleven chapters is written by an American scholar who came to know a particular Muslim well. These chapters are written from a personal perspective because this is the only way we can come to know people.

It is our hope that readers will appreciate the variety of ways each person copes with his or her changing world, which for many contemporary Muslims has meant coping with war, exile, economic dislocation, and the tensions that ensue at the national, community, family, and personal levels. It is also our hope that readers will see how Islam figures in these Muslims' lives, how they renegotiate its meanings in their changing contexts, affirming long-held ways or modifying them. In short, we hope that readers will come to see some of the diversity that is always the hallmark of a world religion and will recall the stories of these individual Muslims at a time when the human face of Muslims has been overshadowed and ignored.

This volume was initially conceived by Linda Walbridge who commissioned many of the articles shortly before her death from breast cancer in 2002. She was dear to us both. We hope that we have been faithful to her conception and that this book will be true to her vision that the anthropological study of culture is ultimately based on one human coming to understand another.

FRANCES TRIX AND JOHN WALBRIDGE
ISTANBUL, TURKEY

Introduction

Frances Trix and John Walbridge

All but three of the people discussed in this book are still alive. You could pick up a phone and telephone each of them. Most have e-mail. Most have read the articles about themselves and commented on them. All of them know that this book is being published. Baba Rexheb, one of the three who are no longer with us, is buried near Detroit. If you take Telegraph Road south from I-94, turn left at Northline Road, and go a few blocks, you will see the copper roof over his tomb on the right. The old baba who is the custodian of the mausoleum will let you in so that you can say a prayer and light a candle. These are all real people, who live in real places, and who eat, sleep, and go about the ordinary business of life.

Though they are Muslims, they are not the faceless, anonymous Muslims of television and the press. Each in his or her way is convinced that Islam is precious and true, something to be passed on intact to another generation. Each, though to varying degrees, has seen his or her life disrupted and changed by modernization and globalization. One died in the aftermath of the American invasion of Iraq.

At the same time, Muslims today also know the earlier history of the great Islamic empires that flourished from the seventh to the eighteenth centuries. Damascus, Baghdad, Cairo, Istanbul, Cordoba, and farther east Shiraz, Bokhara, Samarkand, Lahore, and Delhi were all centers of learning, trade, art, and political power. The lives of the people written about in this book were shaped by the traditions of the great Islamic civilizations of the past, as well as by the disruptive legacy of European colonialism. Thus, to better understand their lives, and the lives of other Muslims today, it is important to take this history into account.

The Living Past

Muslim World before European Colonization

Let us look back then on a precolonial Muslim world. A well-informed Muslim living in 1798, the year that Napoleon invaded Egypt, would have surveyed his world with some satisfaction and confidence. Though Muslims had at times suffered setbacks, Islam had enjoyed steady growth from century to century. The Prophet Muhammad's realm had grown from a small city-state of Medina in the Arabian Peninsula in the 620s to the world's largest empire in a single century. Islam was the most powerful and progressive civilization in the world throughout the Middle Ages. Muslim scholars expounded the teachings of Islam with subtlety and depth. Muslim architects built some of the world's most beautiful buildings, including the Alhambra in Spain, the Suleymaniye Mosque in Istanbul, the Dome of the Rock in Jerusalem, the Royal Square in Isfahan, and the Taj Mahal in India. Muslim artists and artisans created books and household objects—textiles, rugs, inlaid wood, porcelain, tiles, and bronzeware—of exquisite beauty. For five hundred years Muslim scientists were the finest in the world, rightly famous for their knowledge and sophistication in mathematics, astronomy, pharmacology, optics, and medicine. It was a literate and urbane society in which a scholar, artist, or educated man might travel thousands of miles and find appreciative employers in every major city. Arabic, Persian, and Turkish, the languages of scholarship, culture, and administration, linked lands across the Middle East and southern and Central Asia. Books were everywhere.

There had sometimes been defeats. Invading Mongols in the thirteenth century had killed the caliph, the titular leader of Islam, and wrought devastation from Central Asia to Iraq. There had been other barbarian invasions, and conflicts with rival civilized powers. Nevertheless, invading barbarians had usually been converted to Islam and adopted Islamic ways. The European Christians who called themselves Crusaders had eventually been driven out of Syria and Palestine. In the fifteenth century, Islamic territory had been lost forever in Spain, but Islam was still strong across the straits in Morocco and had gained territory in India, China, Southeast Asia, Central Asia, and Africa. There was a powerful Islamic empire administered from what is now Turkey and many lesser but still vital states in Asia and Africa.

To be sure, the European Christians were a growing threat. Their military and naval technology had begun to give them decisive advantages in war. For three hundred years they had increasingly encroached on the traditional trade routes. In 1798, however, it was still possible for Muslims to view the Europeans as no more than a nuisance, rather in the way that an American at the dawn of the twenty-first century might view the long-term threat posed by the revitalization of China and India. All in all, our thoughtful Muslim at the end of the eighteenth century would have foreseen problems but not disaster.

Disruptive Legacy of European Colonialism

But it was disaster that came. European economic and military dominance surged. One by one, the old Muslim states, their traditional economies and their armies wrecked, fell into the hands of European powers. The French took most of Islamic North and West Africa. The British overthrew the Islamic Mogul Empire in India and occupied the Muslim kingdoms of Malaya. With the help of Russia, most of the Balkans broke away from the Ottoman Empire, forming Christian states that expelled or killed most of their Muslim populations. The Dutch took Indonesia, the Russians Central Asia, and the Italians Libya. With the defeat of the Ottoman Turks in World War I, their possessions in the Near East were divided between the French and British. Only a few Muslim states, notably Turkey, Afghanistan, and Iran, retained their independence, mostly by playing on the jealousies of the European states. The former Muslim states were incorporated into European empires, and their colonial masters began programs to modernize their colonial possessions by remolding them on European models. Students who a few years earlier would have studied Arabic or Persian now studied English or French. Islamic and traditional law was replaced by European law. Islamic institutions lost their official support and withered. Economies were reorganized as part of a colonial system feeding foreign factories. The political, economic, social, and psychological disruption was total.

Then, with equally bewildering speed, the Europeans threw away the power they had gained. World War I, an orgy of pointless and hideously expensive mutual bloodletting, bankrupted the great European empires and lost them the moral, economic, and military standing that might have allowed them to hang on to their new possessions. The European claim to represent a superior civilization now rang hollow. Nationalists began agitating for independence in the Muslim parts of the European colonial empires. World War II sealed the doom of the European empires. No longer able to afford their colonies, the European empires granted independence to their former possessions, leaving behind shaky nationalist regimes and societies that had been partly, though by no means completely, Europeanized. A bitter legacy of colonialism for the Islamic world was Israel, a state founded largely by European Jewish refugees who came to Palestine under British protection and supplanted the Arabs, Muslim and Christian, in much of the old British Palestine mandate.

Polarizing Influences of the West

A second wave of contact with the West occurred after World War II during the great conflict between the United States and the Soviet Union known as the Cold War. The newly independent Muslim states found themselves caught up in this conflict. Ambitious statesmen sought to advance their own interests and those of their countries by allying themselves with either the Soviets or the Americans or by playing them off against each other. States like Egypt and Iraq leaned first to one side and then to the other, hoping to maximize political, military, and economic support from one or both superpowers. Islam as a religion seemed irrelevant in this clash of titans.

At the end of the 1970s, Islam abruptly reemerged as a political force. In many countries nationalist parties or dictators had ruled since independence. The nationalists, usually secular and often socialist in orientation, had made great promises that were not kept. Economies, often smothered by aging nationalist oligarchies, remained mired in poverty. The last straw was the 1967 war with Israel in which three leading Arab Muslim states suffered a humiliating total defeat at the hands of Israel. The remaining third of the old Palestine, along with portions of Egypt and Syria, fell into Israeli hands. Disillusionment led many Muslims to turn away from nationalism toward a political form of Islam.

In 1978, an unpopular monarchy in Iran supported by the United States was overthrown by a mass revolution led by students and Muslim clerics. The new regime, the Islamic Republic of Iran, was a sort of theocracy. For the first time in more than two hundred years, Iran, now ruled in the name of Islam, was completely free of foreign influence. This revolution captured the imagination of Muslims across the Islamic world. The political potency of Islam was further demonstrated during the next decade when an American-supported coalition of Islamic guerrilla bands first resisted and then defeated a Soviet occupation in Afghanistan.

The Muslims of This Book in Changing Times

All the people discussed in this book are grounded in the Islam of the past. They never doubt that Islam is true, that it is important and valuable. Islam, if correctly understood and practiced, will provide answers of both spiritual and practical value. On the other hand, the lives of each of these people have been affected or even devastated by the forces unleashed by the European colonial adventure. For some its impact on their lives has been superficial. They have remained grounded in the traditional Islamic life of the past. Baba Rexheb, an Albanian monk, lived almost half his long life in Michigan, but he remained devoted to passing on an old spiritual tradition grounded in Ottoman Turkish culture, albeit under new circumstances. Had he lived five hundred years ago, he would have been much the same person. Fadma Lhacen, a Berber healer and weaver, was touched by the modern world mostly because it drew her children away to the new modern cities. She too would have been much the same had she lived five hundred years ago. For African-Americans, the slave trade that brought them to North America and the Caribbean was a colonial enterprise. Some Muslim African-Americans make the case that they are not converts to Islam but "reverts," since undeniably many slaves from Africa were Muslim before they were taken to America. Even without this connection, religion has held a central place in African-American society, and Anisa was raised in Sunni Muslim tradition by parents for whom Islam continues to be the cornerstone of their lives and family.

Other people in this book have been affected by the globalization and modernization unleashed by colonialism in deeper ways. While some have moved thousands of miles from where they were born, they have been displaced psychologically as well, forced to come to terms with new cultural worlds. Sheikh El-Zein, born and raised in Lebanon, later studied in Egypt and in Iraq. But he ended up as religious leader of a Muslim community in Senegal, in West Africa, far from his homeland both geographically and culturally. And yet both Lebanon and Senegal were colonized by the French. Emine, a college student of Turkish heritage in Berlin, was born in Germany and wants to work as a teacher there, but she encounters professional difficulties due to her wearing of the headscarf. The scarf is basic to her integrity as a practicing Muslim, but it has become increasingly controversial in Germany, thanks to the stand of the French government against the wearing of identifiably religious garb in schools. For such people, the essence of their lives as Muslims is navigating new cultural waters resulting from colonialism, modernization, and globalization.

Finally, there are some written about in this book who have been able to reconcile their Islamic past with the new globalized world. Sabah Naim, a successful young female artist in Egypt, seems to be quite comfortable between her traditional and religious family life and the life of an artist. Salma, a woman in Afghanistan working for an aid organization, balances her professional ambitions and the traditional roles of an Afghan woman.

In short, the past two centuries have brought extraordinary change to the Islamic world, as they have elsewhere. The change, often in the form of political, economic, and social disaster, has perhaps been more traumatic in the Islamic countries than elsewhere, just because Muslims had been so successful in previous centuries. For more than a thousand years, wherever a Muslim lived, chances were that his or her land was ruled by Muslims or soon would be. Muslims were usually richer, more powerful, and more civilized than their neighbors. They were educated and cultured a thousand years ago when most European Christians were little more than barbarians. When Europeans came to Palestine as Crusaders, the Muslims who came in contact with them noted that they smelled bad and had coarse habits. As recently as a century ago, a Muslim in Istanbul, Damascus, or Thessalonica could take satisfaction in being an Ottoman, a subject of the greatest Islamic state in the world. The change was sudden, and was the more painful by contrast with the glories of the Islamic past. All the people in this book have been trying in their various ways to come to terms with the trauma of this fall from power.

Diversity in Muslim Traditions

Though the eleven people in this book have been shaped by the same great forces of history, they are certainly not the same. The tumult of the past two hundred years has affected places differently. Moreover, these people represent different traditions rooted far back in the Islamic past. Finally, these are just people, with the peculiarities of interests, character, and personality of people anywhere. Sabah Naim, the young female visual artist in Egypt, is a typical artist, not a typical Muslim. She stands for herself alone. Similarly, Majid, the Iraqi who lived in Great Britain, was raised in a family of Shi'a Muslim scholars, but the formative fact of his life was political due to the regime of Saddam Hussein.

That said, there is also genuine religious diversity among these eleven people, diversity that is rooted in developments in Islamic life and thought that took place a thousand years ago or more. Some of the Muslims in this book would not approve of the Islam of others and might not even consider it Islamic. At the risk of simplifying a little, we can identify four strains of Islam represented in this book, though some of the people here fall into more than one category: legalism, Sufism or mysticism, Shi'ism, and folk or popular Islam. Except for the Shi'a, who really are a distinct community, these are not denominations in the Christian sense, because the ideal of Islamic unity is too strong.

Legalism

Islam, like Judaism, is a religion of law, not of creed. While a Christian might say that a true Christian is someone who has the correct beliefs about Jesus Christ as his savior, a Muslim would most likely say that a true Muslim is one who prays, fasts, gives alms, and obeys the other laws of Islam. Muslims do share common beliefs, but they have always been much more concerned with details of law than with nuances of belief. Like the rabbis of Judaism, many of the most serious Muslims down the centuries have expressed their faith through the careful study and practice of religious law. Legalistic Islam is preeminently the Islam of books, being based very directly on the Qur'an, on the recorded sayings and practice of the Prophet known as *hadith*, and on the books of the great Islamic religious scholars. The demands of the law of Islam are in the background of many of these stories: Anisa, Emine, and Salma, who choose to cover their hair; Sheikh El-Zein who revived and expanded practice and understanding of Islamic law among Muslims in West Africa; Prof. Koesnoe, the Indonesian legal scholar; and others.

Sufism, Islamic Mysticism

For more than a thousand years, there has been a vibrant tradition of Islamic mysticism coexisting with other forms of Islam. For many centuries, the mystical masters and the great legal scholars shared spiritual authority across the Islamic world. The mystics sought the inner meaning of the laws and teachings of Islam, and practiced a more expressive communal form of worship. They chanted the names of God and spiritual poems in graceful ceremonies to express their love of God. Baba Rexheb, the Albanian dervish, is the most important Sufi in this book, but Baba Shafqat is both a Shi'ite and a Sufi, and the profound spiritual richness of Sufism touched the lives of others in this book.

Shi'ism

When the Prophet Muhammad died, there was a dispute about succession. Most of the community supported the leadership of an old and respected friend of the Prophet, but a minority supported the Prophet's younger cousin and son-in-law, 'Ali. About fifty years later, 'Ali's son Husayn revolted against the corrupt Islamic ruler of the time and was killed with most of his family and supporters. The modern Shi'ites are the sect that coalesced around the Imams, the twelve rightful heirs of the Prophet, and the memory of Husayn's martyrdom. The Shi'ites are distinguished by a fervent, sometimes tormented grief for the sufferings of 'Ali, Husayn, and their descendants and supporters. The atmosphere of Shi'ism is very different from the more optimistic Sunni Islam, the majority school in Islam. There are several, quite diverse, Shi'ites in this book. Baba Shafqat in Pakistan devotes himself to reminding people of the sufferings of Husayn through song and poetry. 'Esmat, a widow in Iran, consoles herself for the tragedies of her life through her prayers to the Imams. Sheikh El-Zein, a scholar trained in Shi'ite law, leads a community of Lebanese Shi'ites. 'Abd al-Majid was the son of the most eminent scholar in the Shi'ite world.

Popular or Folk Islam

Not all Islam is based on books. Much is the custom of the various and far-flung lands of Islam. Each country, province, city, and village has its own customs that its people identify with Islam. Tombs of local Muslim saints are important centers of Islam in each place. Some customs obviously predate Islam or contradict in one way or another the official Islam of the great books of Islamic law. Some are denounced by more sophisticated Muslims as mere superstition. The Berber woman Fadma's healing practices and Baba Shafqat's Shi'ite music are each folk practices that have little direct warrant in the Qur'an but retain Islamic understanding nonetheless. Often, there is conflict between local custom and those who advocate a more standardized form of Islam. Salma, the Afghan aid agency manager, carefully negotiates a line between Afghan custom and her notions of what is appropriate for a Muslim woman.

Such diversity long predates the tumult of the past two centuries, but the conflicts have now become sharper. Muslims of different types and temperaments are thrown together by the ease of modern communications. Islamic communities in the West are more diverse than any that ever existed in Islamic countries. Sufism and folk Islam have come under attack as superstition by both legalists and secularists, but it should be noted that the pious and erudite Prof. Koesnoe spent his career defending local custom as a source of Islamic law.

Of course, there are also Muslims who are Muslim by culture only for whom religion plays no significant role in their daily lives. For many, political or national loyalty takes the place of religion. Nevertheless, as with their Abrahamic cousins, the Jews, even completely nonpracticing Muslims usually retain a residual loyalty to their ancestral religious community, and it is rare for a Muslim to renounce his faith entirely or to convert to another faith.

Diversity in Disruption

These have been the deep Islamic roots of the diversity among our subjects, but there are also differences based on the historical experiences of each. Not every country has had the same experience of the tumult of the modern world, nor has each region, class, or family.

War

When we looked at the chapters written for this book, we were struck by how pervasive a force war had been. Seven of the eleven people discussed had been affected to varying degrees by war. The life of the Afghan Salma has been overshadowed by the thirty years of war that have followed the Communist coup and Russian occupation. Baba Rexheb and Prof. Koesnoe had been involved, for the worse and the better, in World War II and the civil wars that led to major changes in their countries. Esmat Khanum's son fought in the long and bitter Iran-Iraq War. Sheikh El-Zein ministers to a Lebanese immigrant community whose newcomers are refugees from the Lebanese civil war. Majid al-Khu'i's life was shaped by resistance to the regime of Saddam Hussein and ended in the internal fighting after its fall. Baba Shafqat's life in Pakistan has been overshadowed by the conflict with India and the aftereffects of the Afghan civil war.

Other wars loom in the background. Since World War II there have been five wars between Israel and her Arab neighbors, three major wars involving Iraq, several wars between Pakistan and India; separatist revolts in Chechneya in Russia, the southern Philippines, Aceh in Indonesia, western Oman, southern and western Sudan, and Somalia; struggles for independence in Algeria, Western Sahara, Bosnia, Kosova, and Kurdistan; and civil wars in Lebanon, Iraq, Indonesia, Yemen, and Algeria. War and the threat of war have been pervasive in the modern

Muslim experience. The disruption caused by war has been enormous, scarring the lives of individuals, communities, and nations.

Economic Disruption

In the nineteenth century the world economy came together into a single, closely connected whole. Civil war in America brought prosperity to Egyptian cotton farmers and merchants; peace brought depression. The discovery of silver in Colorado and Nevada caused inflation and economic ruin in Iran. And everywhere, the industrial revolution in Europe and America ruined traditional craftsmen who could not compete with the cheap products of European factories. For the most part, the countries of the Islamic world have never quite recovered from this shock. The successful economies have been based on the export of raw materials, mostly oil but also sometimes agricultural products. Except for oil-exporting countries, the Islamic world remains on the fringes of the world economy, never having developed robust industrial economies able to compete with the West or the emerging industrial economies of East Asia.

The effects of this long-term economic transformation and disruption ripple through the lives of almost all the people in this book. Fadma Lhacen's sons have moved to the cities. Emine's family lives a life not quite German, no longer quite Turkish, in Berlin, because that is where there is work.

Cultural Disruption and Emigration

Cultural change and disruption has come from two directions. First, colonialism and now globalization have brought Western culture into Islamic societies. Colonial administrations undermined traditional educational systems and replaced them with systems modeled on those of the home country. Western, and particularly American, television and movies are watched throughout the Islamic world. Western styles of dress, etiquette, consumer goods, and so forth are increasingly the norm throughout the world and in Islamic countries. Since colonial times, European-style districts have grown up alongside older traditional neighborhoods. Perhaps the most pervasive instrument of change has been Western technology. Many welcomed the technology for its practical applications, but for better or worse with new technology have come new ways of behavior and thought.

Second, many Muslims have moved to Western countries. They live a style of life constrained by their adopted countries while trying to retain some remnant of their own ways in their homes and Muslim communities, while their children become increasingly American or European. Freedom is a particularly ambiguous gift. Many Muslim clerics who have settled in the West have found that they have more freedom to practice their religion than they had in their home country. On the other hand, Muslim parents, like all parents in Western societies, worry that their children will lose their way among the temptations of modern Western culture.

Colonization and Occupation

We have mentioned many of the long-term economic and cultural effects of colonization and occupation, but some of the effects were more immediate. During colonial occupations in most of the Islamic world, the colonial rulers simply destroyed many indigenous institutions by force or by withdrawal of official support. Those who resisted occupation were jailed or killed. Political resistance was crushed by superior military technology. Most of all, there is a lingering bitterness, based on the humiliation of living under foreign rule.

The Unity of Islam and the Individuality of Muslims

Muslims everywhere affirm the Oneness of God, the centrality of the Qur'an as God's Message, respect for the Prophet Muhammad, on whom be peace, as the messenger of God, and respect for his family. There is shared adherence to the practices of declaration of faith, prayer, alms-giving, fasting, and pilgrimage, although as with all world religions, there are differences of emphasis and expression. The very greatness of God with respect to humankind renders differences among people inconsequential and leads to a pervasive understanding of the equality of all people before God.

The unity of Islam has been one of its tenets. And in fact, there is more unity in Islam than in most world religions. But at the same time, there have been divisions among Muslims since the earliest years of the faith. The oldest and deepest is that between Shi'ites and Sunnis. There is a prophecy attributed to the Prophet predicting that Islam would be divided into seventy-two (or seventy-three) sects, only one of which would attain salvation, and medieval Islamic theologians attempted to catalog these "heretical" sects.

Nevertheless, most religious divisions among Muslims have not resulted in the rise of separate denominations like those in Protestant

Christianity. There are four distinct schools of Sunni Islamic law, but the members of the four schools have almost always accepted each others' legitimacy as Muslims. Even Sunnis and Shi'ites have generally accepted that members of the rival sect are Muslims, albeit misguided in certain respects. When confronted with the fact that Muslims are divided in certain ways, most Muslims will insist that these divisions are superficial and do not affect the fundamental unity of the Islamic nation. Division is often explained as the result of Muslims failing to live up to the demands of their faith.

The ideal of unity of all Muslims continues to have powerful ramifications in the modern world. An attack on a Muslim country by a non-Muslim nation is resented by Muslims across the Islamic world. The annual pilgrimage to Mecca, now made by more than two million people each year, reinforces the ideal of Islamic unity, as does the experience of Muslims in the West, who experience the diversity of the Islamic world in their mosques and Muslim organizations, and who increasingly work together to present Islam to the larger non-Muslim societies in which they live. As it happens, most of the people in this book have had personal experience of the larger world—both of the diversity of the Islamic world and of the larger non-Muslim world beyond—so for them the ideal of Islamic unity is likely to be a practical necessity.

That said, these eleven people are at the same time individuals, shaped by many things other than their religion, and not least by their personalities. Probably, if you asked them, each would say that the most important thing about them is that they are Muslims. Yet their situations and conditions also speak to commonalities that transcend religion. When Esmat Khanum prayed to the Imams for the safety of her soldier son, she was a frightened loving mother whose child was in danger and beyond her help. Three of our people are artists of one kind or another: a visual artist, a weaver, and a singer. They are Muslims and, to be sure, their religion is interwoven with their art, but as artists they resemble other artists the world over. Six count teaching as an important part of their work, either at the community level or within colleges or universities. Like all teachers, to be effective they must know their students and be able to reach out to them. Several are scholars whose study touches Islam, but like scholars the world round, their passion is to learn and understand. Finally, two are healers, and the obligations and ideals of healers and physicians are older than those of Islam. But at the same time, for both Dr. Cook, the pediatrician, and Fadma Lhacen, the traditional healer, there is the pervasive understanding of God working through them in their healing.

These are, to be sure, special people, people who deeply touched the authors of these chapters, so much so that they wanted others to know of them and their lives. They are special people because of their characters and personalities, and because of the important role Islam has played in their lives.

Most of all, they are people worth knowing.

A Note on the Grouping of Chapters

These chapters could have been arranged in many ways. We have chosen to separate them into three categories that we think reflect the relationships of our people to their cultural surroundings.

Grounded

Four of the people discussed in this volume, Baba Rexheb, Fadma Lhacen, Baba Shafqat, and Esmat Khanum, seem to us to have remained grounded in traditional culture. Though in every case the modern world has touched their lives, they have been able to continue to draw on an older form of Islamic culture to make their ways in the world. They would, we might say, have been much the same people had they lived several centuries ago.

In a New Place

For Emine, Sheikh El-Zein, and Majid al-Khu'i, the experience of moving to a new country has reshaped their lives, and in one sense or another the task of making their way in a new cultural setting has defined their lives. In these cases, it seems to us that their lives have contained an unresolved tension.

Reconciled

Finally, Salma, Dr. Anisa Cook, Dr. Haji Mohamad Koesnoe, and Sabah Naim seem to us to have faced a new cultural reality and been able to transcend it. They seem to have faced the cultural changes and disruptions of the modern world and found their own accommodation with it.

Map Homelands of Muslims in This Book (copyright John Walbridge, 2007)

PART I

Grounded

CHAPTER 1

"The Seeing of Our Eyes:" An Albanian Sufi Baba *(Balkan Sufi Leader in United States)*

Frances Trix

Baba Rexheb liked corn flakes for breakfast. He would get up early in the morning, pray, and then go down the stairs in his slippers to the basement kitchen of the *tekke* (a sort of Muslim monastery). In the early years when the *tekke* had only recently been established by Albanian Muslims in America, the kitchen was a low-ceilinged room with a narrow table, long enough for twenty people to sit at comfortably. The table was covered with red vinyl that was nailed down for ease of cleaning. Baba always sat at the head of the table, for he was the head of the *tekke*. To his back was the large squat twelve-burner stove whose burners would be covered with pots of food during the holidays, or at times of engagement parties or memorial services. But in the early morning Baba would sit by himself at the table and eat his corn flakes and milk, careful afterward to wipe his long gray beard of any milk.

Although Baba Rexheb (pronounced "re-jeb") passed away in 1995, to this day when I see corn flakes, I think of Baba in his quiet *tekke* kitchen in the morning. As for Baba himself, I think corn flakes reminded him of the bread made from corn that shepherds and other poor folk ate in his homeland of Albania. People in towns ate bread made from wheat, and those in cities in good times had both wheat bread and rice, which was even more expensive in the small mountainous country of Albania, which lies north of Greece on the western side of the Balkan Peninsula. But when Baba came to America in the early 1950s to found a Bektashi *tekke*—that is, a center of the Bektashi Muslim order, a seven-hundred-year-old Sufi or Muslim mystic order—some Albanian Muslims in America were wary.

Would his Bektashi *tekke* divide the Albanian Muslim community that had only recently founded a Sunni or orthodox Muslim mosque in 1949? If Albanian Muslims in America supported Baba, would their relatives back in Communist Albania suffer? Despite these concerns, there were other Albanian Muslims in America who recognized the importance of having a Bektashi center here. They pooled their resources and bought an eighteen-acre farm on the outskirts of Detroit to serve as the first Bektashi *tekke* in America. Gradually, many of those Albanian Muslims who had initially been afraid and stayed away came around. One way that Baba welcomed them was by serving them bread made from corn in the basement kitchen of the *tekke*. This was a simple yet profound way of saying to them that all the divisions and fears of recent times are put aside; what matters is that we are all Albanians and that we have not forgotten the simple ways of our homeland.

All this I learned later, for I studied with Baba at his *tekke* for twenty-five years. But when I was just getting to know Baba, his eating corn flakes for breakfast made him seem more like me, that is, American.

Figure 1 Author, Baba Rexheb, Dr. Bedri Noyan, Dervish Arshi

Baba proudly did become an American citizen. And he often referred to America in Albanian as *ky vend i bekuar* (this blessed land).

America was held in high esteem by Albanians of all faiths. They recalled how President Wilson had stood up for the small countries in Europe after World War I. But Baba's love of America was based on more recent events. During World War II, Albania had been overrun by Italian and German armies, and with its fragile government destroyed, civil war had broken out, from which the Communists rose to power. Albania had been the only country with a Muslim majority in Europe in modern times—70 percent of its people were Muslim, 20 percent Orthodox Christian, and 10 percent Roman Catholic. Communist ideology denigrated all traditional religious beliefs and practices. But the Communists were particularly hostile to the Bektashi Muslim and Roman Catholic religious leaders—the religious leaders who had commanded special love and respect among the Albanian people. Thus, Baba had been forced to flee Albania in 1944 because of his status as a major Bektashi religious leader and one who had spoken out publicly against atheist communism. The freedom of religion in America thus meant more to Baba than to many of us Americans who took it for granted.

Baba always made sure that Thanksgiving, a special American holiday, was celebrated at the *tekke* in Michigan. On Thanksgiving Day, the cook at the *tekke* would make turkey with stuffing and gravy. Later, when Baba found out that my young son liked turkey, he made sure that the cook at the *tekke* made turkey every week on the day I came to study with him. That way we had turkey for the midday meal at the *tekke*, and I was always given meat to take home to my son. This sort of generosity, this reaching across cultures, touches me more today as I think back on it. And yet, I think it was this sort of practice, this reaching out and respecting local culture and people, that characterized the Bektashis in history and made them so successful as missionaries of Islam.

The Bektashi order to which Baba belonged was founded in the thirteenth and fourteenth centuries in central Anatolia, in what is today Turkey. The founder, Haji Bektash Veli, was a Turkic holy man from Khorasan, now northeast Iran, who traveled westward to Anatolia, as did many Turkic peoples at that time, spurred on by Mongol armies at their backs. The Mevlevi order, sometimes referred to as the "whirling dervishes," was also founded in Anatolia in the same period. Both orders are Sufi orders; that is, they are mystic orders of Islam whose members seek to come closer to God through special ways of praising God. For the Mevlevis, this praise takes the form of graceful turning, pivoting with one hand raised above the head and the other outstretched, hand pointing down, to signify that all comes from God to earth. For the Bektashis, this praise of God takes the form of chanting religious poetry, alternating with the sharing of ritual food and drink.

Like other Sufi orders, for there are many in Islam, the Bektashis have clerics who live in their *tekkes*, namely the dervishes or monks, and the babas or abbots. Bektashis also have initiated lay members known as *muhip*, who live in their own homes but come to the *tekke* for private ceremonies and for fellowship, as well as members known as *ashik*, who are not initiated but are drawn directly to the *tekke* by their love for the baba. Finally there are families whose current relatives or ancestors were affiliated, and so they consider themselves Bektashi. What is unusual about the Bektashis among other Sufi orders is that they have always accepted women as initiated members. The important role of Kadincik Ana, the woman who was the first person to recognize and support Haji Bektash Veli in Anatolia in the thirteenth century, is clear in the earliest Bektashi texts. Thus, the fact that I am a woman and had studied with Baba for many years is not so strange.

Another remarkable quality of the Bektashis is their openness and tolerance for other religions. My own personal experience as an American woman of Christian background whom Baba taught for many years is testimony to this. But it was also clear in smaller gestures, like the way Baba always telephoned both my parents and me on Christian holidays to wish us well. Albanian Christian leaders were always invited to the celebrations of Bektashi holidays, as was Baba invited to their churches for their holidays, which reflects the commitment of Albanians not to let religion divide them. Theologically, it could be argued that the Bektashi veneration of holy leaders and tombs of saints, and their intercessory beliefs and practices, parallels Christian veneration of saints. This does, however, set Bektashis apart from Sunni or orthodox Muslims, many of whom do not believe in intercession through saints or religious leaders.

What further sets Bektashis apart from Sunni Muslims is the gentle humor with which they criticized the rigidity that Sunnis sometimes brought to their practices and beliefs, such as praying five times a day or insisting on the desire for heaven and fear of hell as motivators of good conduct. In contrast, Bektashis pray twice daily, in the morning and evening, but in a more spiritual sense they pray constantly through lives dedicated to God. As for heaven and hell, Bektashis refuse to see these in terms of simplistic moralizing of reward or punishment. Rather, what motivates Bektashis is their desire to come closer to God, to move toward the presence of God.

The lack of rigidity in Bektashi practices and their respect and openness to other cultures and peoples made them especially effective missionaries of Islam among the largely Christian peoples of the Balkans. The Bektashis became chaplains to the special Ottoman troops, known as the Janissaries, who took over much of the Balkans in the thirteenth, fourteenth, and fifteenth centuries. Many of these troops were former Christians, and the Bektashi mediation of Islam suited them well. The Bektashi dervishes and babas settled among the Christian peoples there. It was often the character and spiritual qualities of particular babas that drew local people, just as I was drawn to Baba Rexheb in Michigan.

First Meeting With Baba Rexheb

When I first met Baba Rexheb I was twenty years old. I knew very little about Islam and nothing about Albania. From the point of view of the Albanian Bektashi community, I was distinguished by my ignorance, although they were much too polite to even suggest this. I had telephoned to see if I could visit this unusual center. I was told that the door was always open, for hospitality is a basic tenet of both Bektashi practice and Albanian culture. I was ushered into the main room where Baba was seated, with one leg curled under him, in a corner armchair. I marched right across the carpet and reached out my hand to shake his hand.

Had I been more observant, I would have noticed that everyone else had taken their shoes off at the door. This was a sign of respect but also practical, for the *tekke* was a working farm with many animals and mud outside. As for greeting Baba, I did not know that the usual way was for a person to reach out and take Baba's hand, kiss the back of it, and then bow one's head so that the forehead would touch the back of Baba's hand. This was also expected at leave-taking.

What I do remember of this first encounter was Baba's serious expression and his headgear—a short, white cylinder made of felt with a green band around its base. A much larger replica of this headpiece was found on the roof of the *tekke*. But I don't remember what we talked about then. Baba's English wasn't that strong and I knew no Albanian. Soon the dervish brought in cups of coffee. Historically, coffee spread north from Ethiopia and Yemen through the Sufi centers where dervishes drank the bitter drink to help them stay awake at night during ceremonies of praising God. From Sufi center to Sufi center, coffee finally reached Istanbul, the capital city of the Ottoman Empire. There an argument arose as to whether coffee should be considered an alcoholic drink and therefore proscribed by the Qur'an. But coffeehouses were already becoming popular in Istanbul, and the very popularity of the drink overrode the controversy. From Istanbul, coffee spread to Vienna and then to western Europe.

After the coffee was served, a box of chocolates was passed around as a sign that the visit was over. But before I stood up to leave, another baba with a long white beard appeared and said something to Baba Rexheb in Turkish. I had begun studying Turkish that year and so I understood him. Amazed, I turned to Baba and asked in Turkish, "Do you speak Turkish?" He answered me in Turkish and said that he did. In fact, most Bektashi poetry is in Turkish, so it was not at all unusual for Baba to know the language. Further, the Ottoman Turks had ruled Albania for five hundred years, and Baba had gone to elementary school in Turkish in Albania before Albania declared its independence in 1912.

What this meant for me was that Baba and I had a common language, or at least the basis for one. Immediately I realized that I wanted to study with him, but somehow I also knew that I couldn't just ask. It took three or four more awkward visits to the *tekke* until Baba Bayram, who had first spoken in Turkish and who turned out to be the cook for the *tekke*, understood that I wanted to study with Baba Rexheb. He suggested it and mercifully Baba Rexheb agreed. I did not know then that I would continue to come to the *tekke* and study with Baba for twenty-five years. But I did know from the outset that Baba had a dimension that my professors did not have and that I was drawn to. At that time, I was studying language, literature, and religion at the university from professors who had studied these fields. But Baba lived them.

Baba in Community

In my weekly visits to the *tekke*, I read Bektashi poetry in Turkish with Baba and learned about the order and its beliefs, but I also came to know other Albanian Muslim community members and how the *tekke* was a part of their lives. Without my noticing it, the *tekke* slowly became a part of my life. At some point, I too learned to kiss the back of Baba's hand and draw my forehead to it in greeting and leave-taking. But I have no recollection when this began; it was too gradual.

After I had been studying with Baba for ten years, my friends and relatives would ask me, what do you learn in that place? Despite the cultural richness of the *tekke*, and the erudition of Baba, for he was at home in Arabic, Persian, Ottoman Turkish, and centuries of Islamic sacred texts, as well as in Greek and Italian, I never quite knew how to answer them. Much later I realized that I was learning a relationship, and that the relation between Bektashi teacher and student is a profound one, unlike any relationship I had had in my many years of schooling. But certain encounters at the *tekke* do stand out.

I remember one Saturday morning in the early years of my study with Baba when we were seated outside on a wooden bench under an oak tree, across from the chicken barn. It was a warm spring day and we were enjoying being outdoors again after the long winter. A station wagon drove up and several Albanian men I had often seen at the *tekke* got out and came over to greet Baba and sit together to his right. About ten minutes later, another car drove up and another Albanian man, along with an American man, got out. The Albanian introduced the American to Baba as a lawyer, and then he proceeded to sit to Baba's left. I asked Baba what was going on.

Baba explained to me in Turkish that two Albanians had had a partnership in a restaurant and they wanted to dissolve it; that morning they were going to decide on what would be a fair division. He also made a slightly disparaging comment about lawyers to me in Turkish; he made the point that they were not needed here. The American lawyer turned to me and asked with a smile what Baba had said. He knew neither Albanian nor Turkish, and looked somewhat befuddled by the whole situation and wondered what role the bearded man in the white hat, long vest, and baggy pants would play. I just nodded to him. What could I say?

Baba then took the partner from the first group of men into the *tekke* for about ten minutes. When he returned, he took the other partner, the man who had come with the lawyer, into the *tekke*, leaving me to make small talk with the lawyer outside. After about ten minutes Baba came out of the *tekke* again and when both parties were seated, announced in Albanian how the partnership would be split. And it was done. They took their formal leave of Baba, kissing the back of his hand and drawing it to their foreheads, and departed. The lawyer looked even more confused, but Baba had been right, his presence was not required. Baba's stature in the community, his integrity, and his fairness were beyond question.

This event stood out in my memory, but what was much more common was for people to come to Baba seeking his blessings. People sought blessings for their children, for good marriages, for their relatives who were ill, for the success of a business venture, for safety on a long trip, and for the health of relatives back in Albania. Baba would take the person to be blessed, or the person requesting the blessing, to a quiet place in the *tekke*, place his hands on the person's head, and pray intensely the old prayers in Arabic, Turkish, and some prayers that he had translated into Albanian. Baba later explained to me that he was passing on their requests, and if *Cenabi Hakk*, that is the Majesty of Truth, the Bektashi term for God, chose to grant them, it would be good, but it was not in his hands.

Many people recognized Baba's authority when he prayed, but there was much humility here too. Whenever I thought of Baba's humility, I would remember the story of the ant going on pilgrimage. Baba would tell me this story when I asked how people could aspire for spiritual perfection. It seemed that an ant was going on pilgrimage to Mecca. People laughed when they heard this. "You, on pilgrimage?" But the ant replied, "As much as *Cenabi Hakk* gives me strength, that far will I go."

Usually the people who came to the *tekke* to seek blessings were adults. But sometimes teenagers came too. I especially remember seeing a young man come with his older sister one afternoon. They arrived in a bright red car, and the young man was clearly pleased with the car. The sister came into the *tekke* and stayed talking with me while Baba went outside to the young man. A few minutes later, much to my surprise, through the window I saw the red car leave the *tekke* with the young man driving and Baba in the passenger seat. What was going on? Fifteen minutes later the car returned and the sister and brother drove off together. When I asked Baba what had happened, he laughed and said that it was the young man's first car.

But the fuller explanation came later. After one of the holiday celebrations, I found myself talking with another teenager, a young man whose family had long been Bektashi, but who himself was born and brought up in Michigan. We were talking about Baba, and the young man mentioned how Baba wasn't judgmental like the parents of so many of his friends. Indeed, immigrant parents are often worried about how their children will fare in the new society of America, and this can often lead to critical judgments. I too had noticed during my studies with Baba that he didn't dwell on my confusions, and at most told me stories or parables when things got rough. I couldn't recall a single direct correction, let alone an admonishment. And then the young man told me how Baba had blessed his car.

When he received his driving license, his parents had bought him a car, but they insisted that he come to the *tekke* to have Baba bless the car. The young man said he thought Baba would take him into the *meydan*, the private ceremonial room, and say a prayer there and then it would be done. But no, Baba asked him to take him to the car.

Then the young man said he thought Baba would say a prayer over the car and then it would be done. But no, Baba told him to take him for a ride. So the young man helped Baba into the car, fastened his seat belt, and took him for a ride. Then he smiled at me. That was it. That was the blessing. No special words, just Baba's trust in him as a driver, along with Baba's presence in his car.

What was particularly Bektashi to me in this was the change that Baba had thereby effected. It was the parents who had insisted on the blessing, not the young man. He had complied to come to the *tekke* because he was a good son and probably because his parents were paying the insurance. But Baba made no reference to the parents. Rather, in trusting the young man to drive him, Baba was instead focusing on a relationship with the young man himself. That words were unnecessary was also astute, for it played against the notion that blessing was simply words.

My young son also recognized Baba as someone to be trusted. It was Baba who had named him. Before he was born, I had asked Baba for a name that was used by both Christians and Muslims. Three weeks after he was born, Baba had come to our home and, praying over him, gently whispered the name, "Ramsay," into his tiny ear. But of course, this was something appreciated by me; what my son experienced was how he was treated at the *tekke*. It was a wonderful place for a child, with all the staircases and the animals, and the people who were working would always stop to make time for a child. Albanian culture is family centered, and children are seen as gifts from God. And there is a Bektashi sense that we are all children before God. In any case, my son loved to come to the *tekke* and would run in, forgetting to take off his shoes. He would jump up on the lap of Baba. Albanian children were better behaved and more restrained. But Baba didn't mind. And as my son grew older, he made thoughtful observations. For example, he noted when he was eight that when I was sick, the priest never telephoned us, but Baba always did.

I had wondered about Baba's own childhood and how he had decided to become a Bektashi cleric. I learned about it inadvertently when I asked Baba how he got his name, Rexheb. Baba sat back, took out his *tasbih* (prayer beads), and told me how his parents had been married for seven years but still had no children. His maternal grandmother then went to her brother, who was a Bektashi baba, and asked him to pray that a child be born to her daughter. He agreed, but said that the child should be for them too—that is, for the Bektashis. She agreed, and soon after her daughter was pregnant. In nine months, in the Muslim month of Rejeb, a boy was born. His mother's brother, that is, his maternal uncle, who was also a Bektashi baba, wrote a poem about his nephew's birth and how he was named after the month he was born in. Due to the earlier words of the child's maternal great uncle, it was known that he would become a Bektashi cleric. Baba said that he took pleasure in knowing this himself. At sixteen, he left home to live in the Bektashi *tekke* and continue his studies with tutors and his maternal uncle, who became his spiritual teacher, as had his uncle's maternal uncle a generation earlier. At twenty-one, he passed his examinations in Islamic studies and took the vows to be a dervish. But who could have guessed then that in a little over twenty years he would be forced to leave his homeland, travel west over the ocean, and establish a Bektashi *tekke* in America?

And who could have guessed that my study of Turkish would make it possible for me to study with someone like Baba? But I always thanked the Albanian Muslim immigrants in Michigan too, for if they had not come earlier to Detroit, Baba would not have come to establish a *tekke* for them. At the same time, Baba was always there for other people too. Albanians of different religions, Sunni Muslim and Roman Catholic, would call on him for prayers. Sometimes if they lived in New York and could not come to Detroit, he would pray for them over the telephone, and then light a candle for them in the private ceremonial room later in the day. Non-Albanians sought him out as well, and he had good relations with the shopkeepers and farmers near the *tekke*.

When Baba first came to Michigan in the 1950s, there was an Italian family that lived across the street from the *tekke*. Baba had spent four years in displaced persons camps in Italy after he was forced to flee Albania, and so his Italian was good. An older member of the neighboring Italian family killed himself; this was in the 1950s and the local Catholic church refused to conduct a funeral for a suicide victim. So the family requested Baba to conduct the funeral. The Bektashi attitude is that if a man kills himself, he is clearly out of his mind. This is an occasion for sadness, not condemnation. Baba wrote a moving ecumenical funeral service in Italian and conducted it with dignity for the family.

Much later, in the 1990s, I also recall an occasion when another non-Albanian requested Baba's assistance. Again it was a local man, one who had plowed snow from the parking area of the *tekke* for a monthly fee for several years. I had reached the *tekke* on a Thursday morning to study with Baba and found him seated with this middle-aged American worker in the main room. Baba told me to ask the man what the man wanted.

The man explained to me that he needed money for the monthly payment on his plow truck. This certainly was not a new need, and the man offered no explanation. I surmised that he had perhaps drunk up the money and now was in a jam, for he made his living plowing snow for people. I explained in Turkish to Baba what the man wanted and why. Baba nodded and immediately reached into the pocket of his long vest where he kept money. I reached across to stop him, suggesting in Turkish that we first write out a note for repayment. But Baba refused, and instead directly handed over to the man the four hundred dollars that he had requested. The man was most grateful and left. I asked Baba, "Do you think he will repay you?" "Perhaps," he said, but it was clear that repayment was not of the essence.

Baba's Passing from This World

During my many years of study with Baba, I sometimes gave talks at academic conferences on Bektashism or aspects of Albanian history and religion. I always wrote out the talks beforehand and then consulted with Baba, for I did not want to misrepresent his order. I had so much to learn. Mystic knowledge in any religion takes both time and spiritual maturity to begin to understand. But in my many years of study, I had at least become better at asking questions.

I recall the time when I finally had the sense to ask Baba, "What is the essence of Bektashism?" Baba answered, "We would never pull the veil from anyone's face." Notice the absence of laws or pillars or commandments, but rather the importance of maintaining the dignity of others. And then Baba told the story of 'Ali, the son-in-law of the Prophet Muhammad and revealer of the spiritual understanding of the Qur'an, who was walking one day along a narrow road in a town. The houses were all made of stone, with high stone walls enclosing them and hiding their courtyards from view. But in one wall, a large stone had fallen out, and as 'Ali passed by he could see that a man and a woman were doing something that they shouldn't have been doing inside the courtyard. So he picked up the stone that had fallen out of the wall and quietly restored it to its place.

When I had asked again about the essence of Bektashism at another time, Baba told the story of some villagers who were en route to the city to give false testimony for payment. On the way, the villagers stopped at a Bektashi *tekke* and the Bektashi baba immediately surmised their mission, but he welcomed them in without saying anything. Then, over coffee and sweets, he proceeded to tell them a story about people in a northern town who had gone to a Bektashi *tekke* for an initiation ceremony of a man as a new *muhip*. At the initiation ceremony, another member recognized the new man as one who had falsely sold him a sick animal in the past. After the ceremony, in the fellowship that followed, the man who had unknowingly bought the sick animal tried to bring up the incident in front of the baba. The baba understood but changed the topic quickly. Again the man tried to bring up how the new man sold him a sick animal, and again the baba saw to it that this message did not get out to the gathering. When the man tried again for a third time, the baba finally said, "Such talk and accusations have no place in a *tekke*." And the villagers understood from this story that the baba knew why they were going to the city. They thanked the baba for his hospitality and returned directly to their village.

So gentle yet potent were the ways of teaching—there was no way I could do justice to them at academic conferences. In any case, before I undertook presentations at such conferences, I would call Baba for reassurance. He would make me smile and usually end by telling me, *bu da geçer ya Hû*; that is, "this too will pass, O He." The "He" referred to is a Sufi reference to God. Indeed, the academic conferences did pass. But more importantly, Baba was getting older and we in his community were not at all prepared for his passing. *Nazarim*, people addressed him, that is, "the seeing of my eyes."

Baba was born in 1901 in Albania, came to America in 1952, and was in his nineties by the 1990s. He had always insisted that he wanted to be buried on *tekke* grounds. The local municipality was not used to such requests, and at first they turned us down, even when we explained that it was an age-old Bektashi custom that the founder of a *tekke* be buried on the grounds of that *tekke*. But then we got a good lawyer, an Italian American whom Baba liked, and this lawyer got the city to give us permission to bury Baba on *tekke* grounds. Plans for his *turbe* (mausoleum) were drawn up and pasted above the head table in the public ceremonial room. The *turbe* would be the traditional octagonal structure with a fountain in front and surrounded by trees and flowers. Inside would be Baba's tomb where people could go and light candles and pray.

It gave Baba peace to be able to look out of his bedroom window and see the building of his own mausoleum. Baba was deeply concerned about who would lead the *tekke* after him, but he knew that by being buried on *tekke* grounds the land would become sacred and could not be sold. When the right leader came, there would be a tekke for him to lead. As for death, he had no fear of it. When a baba passes away, he does not die, but "changes life," or "passes from this world."

When Baba became sick for the last time, he knew it to be the last. And he let me know in the hospital when he wanted to be brought back to the *tekke*. The doctors wanted to keep him at the hospital to do more tests, so they could tell us how much longer he would live. But I told them that we would not believe them and that it was time to return. I knew that Baba would live through the last Bektashi holiday of that year, and indeed he conducted the ceremony in the private ceremonial room from a wheel chair, although he had not been out of bed for weeks. People would bring him special food, including watermelon that he had always liked. But I think he ate only to please us, for I wasn't sure he could taste anymore, and food was not a concern. In his last weeks, I asked Baba again about his life and enquired what was most important to him. "To tell the world about Bektashism," he said. In telling others about Baba, I hope I am furthering that wish of Baba.

Eyvallah

CHAPTER 2

Fadma Lhacen: Healer of Women and Weaver of Textiles
(Moroccan Berber Healer)

Cynthia Becker

In the Tafilalet oasis of southeastern Morocco, one of the first people I met was a seventy-year-old Berber woman, Fadma Lhacen. I did not know then how much I would learn from her. Nor could I know then that her mud-brick home with its largely female household would become my home two years later when I returned to study Berber art and the role of women in artistic production. Instead, when I first met Fadma, we did not even have a common language. Then I knew neither Arabic nor Berber, her native language. I tried to communicate with her youngest daughter in French, but since she had last used French in school eight years ago, it was rusty. We did have a good time though, laughing at our inability to talk and communicating with each other in sign language.

Our laughter was quickly muted when an elderly male friend of the family came on a visit. Fadma Lhacen quickly snatched the deep blue indigo-dyed veil, called a *tahruyt*, and draped it over her head and body. I was surprised to see that once everyone was settled, Fadma Lhacen began to laugh and talk with the male visitor. It seemed like a contradiction to use a cloth to create a physical boundary between two people when a social barrier did not seem to exist. As the conversation swirled around me, I tried to take in all the details of this new environment: the bluish green tattoos on Fadma Lhacen's forehead and chin, the loom set up in the corner of the house, the handwoven carpets scattered throughout the room, and the women's silver bracelets.

Despite the short time I had been in Morocco, I noticed that the distinctive dress of the Berber women, especially in rural areas like the oasis of Tafilalet, stood out. The indigo-dyed blue veil that Fadma wore was embroidered with brightly colored geometric and vegetal motifs and shiny metallic sequins. This was in contrast to the attire of the Arab women of the region, who wore voluminous black cloth coverings (called a *lizar*) that completely concealed their bodies from head to foot. Also unlike the Arab women, Berber women of Fadma's group tattooed their faces and wrists with geometric designs. In addition, these Berber women typically wore silver jewelry, rather than the gold jewelry favored by the Arab women. Finally, Berber women were known for the brightly colored carpets that they wove for their homes.

These artistic distinctions are only some of the cultural features that distinguish Arabs from Berbers in southeastern Morocco. Language is another main marker of the difference between Arabs and Berbers. Berbers, as indigenous inhabitants of North Africa, speak Berber languages that differ from the official language of Morocco, which is Arabic. Fadma Lhacen is fluent in Tamazight, one of three Berber languages spoken in Morocco, as well as in Arabic. Although the term Berber is commonly used by outsiders to refer to the indigenous inhabitants of North Africa, "Berber" is a pejorative term that comes from the Latin *barbarus* and was used by the ancient Romans to describe anyone who was not Roman, and hence a "barbarian." In her native tongue, Fadma does not use the term "Berber" but refers to herself as a member of Ait Khabbash, her particular Berber group.

Many Berbers in Morocco prefer the term Imazighen, which means "the free people." Berbers are scattered across North Africa, from the Siwa oasis in Egypt on the east to southern Morocco in the west. They were probably once the main people of all North Africa, but their numbers now are relatively small. Among all North African countries, the Berber population in Morocco is the largest; approximately 50 percent of Morocco's population of thirty-one million is Berber.

The coming of Arabs to North Africa in the seventh century profoundly influenced the religion, culture, and language of the Berber peoples. It is difficult to know what the religious beliefs of the Berbers before the arrival of Islam were, since the local peoples gradually adopted Islam and learned to pray in Arabic. Most Berbers today speak both Arabic and their particular Berber language. Although some rural Berber women do not know how to read, write, or even speak Arabic, most know how to pray in the language of the Qur'an. Indeed Fadma decided to allow me to stay in her home because of her strong adherence to Islam.

When I returned to Morocco for longer study, I rented a small room on the second floor of Fadma's house. Her daughter later told me that the family had been hesitant about having a foreigner under their roof, but that Fadma had insisted they help me. She reminded them that Islam required that strangers be treated with respect and assisted in any way possible. Fadma later told me that she could not imagine a young, unmarried woman living alone, without the support of her male relatives. Living alone, she explained, would have made me vulnerable to advances from men.

Fadma's husband had died more than twenty years ago, so she ran the household, and no one disputed her decision to accept my presence. Although she had given birth to ten children, some had died at childbirth and others had moved away as adults. When I lived with her, the house was full with Fadma's two divorced daughters, one unmarried daughter, one granddaughter, and one grandson. The house had been built by her sons ten years ago on private land outside a *qsar* (walled mud-brick village). Although she had three sons, none of them lived with her on a permanent basis. The absence of men was common in this economically depressed region of Morocco, since most men had to leave the area to find employment.

All her sons lived in larger cities pursuing their careers, so it was women who ran Fadma's household. When Fadma's sons visited, they gravitated to the long narrow guest room that directly faced the front door, allowing them to greet any male visitors without disturbing the women of the family. This room was the most elaborately decorated room in the house. Long cushions elevated on wooden frames lined its four walls. It also had beautifully carved wooden tables and a large plush carpet, hand woven by Fadma, on its floor. The rest of the house belonged to Fadma and her daughters. Women moved from room to room together, depending on the weather and the time of the day. In the mornings, women frequently gathered in the large kitchen, located to the right of the men's guest room, to prepare meals. In the afternoon, they often sat in the large courtyard to the left of the men's guest room. A portion of the courtyard was in the open air, allowing women to catch the cool afternoon breeze. Three rooms surrounded the courtyard. In the winter, women congregated inside one of the interior rooms to keep warm. Since Fadma found it uncomfortable to use elevated cushions, like those decorating the men's guest room, she and her daughters sat on handwoven carpets covering the floor. I was the only member of the household with a separate bedroom. Fadma explained that when her eldest son married, he purchased a bedroom set and kept it in the bedroom that he shared with his wife. When he moved, he took the set with him. Since Fadma's other sons were unmarried, they, like Fadma and her daughters, slept on the floor and used handwoven carpets as bedding. Individual space did not exist in this communal house. Fadma's sons and grandson slept together in one of the three rooms surrounding the courtyard. Fadma and her daughters slept in another. During summer nights, the entire family congregated to the roof of the house, eating, watching television, and sleeping under the stars.

Fadma's Healing and *Baraka*

The house would have been crowded already with just the women and children who lived there, but it was usually bursting with female visitors as well, many of whom traveled great distances in hopes of being healed by Fadma. She had many techniques for healing, but most often she used herbs to heal illnesses. A woman would explain her ailment to Fadma, who would offer a particular plant to cure the illness. Henna was a favorite, since it is connected to *baraka* (divine blessing). Fadma explained that henna infused the body with positive, healing energy.

Henna is a small shrub with whitish colored flowers that grows in many parts of Morocco, especially the Tafilalet. The leaves of the plant are dried and then pounded with a mortar and pestle. The resulting powder is mixed with water to make a thick green paste. Fadma explained that anything naturally given by God and which comes from the earth has *baraka*. Henna, wheat, and wool are all substances with *baraka* because of their association with the fertility of the earth.

Henna paste is typically applied to the soles and sides of the feet and the palms of the hands and left to dry. Once it is dry it begins to flake off, but women seal it to the skin by dabbing the henna paste with a sugar water and lemon juice mixture. Henna dyes skin a color ranging from reddish brown to bright orange. The color depends on the quality of the henna, the amount of time the henna paste is left on the hands and feet, and the texture of the skin (henna adheres best to roughly textured soles and palms). Good henna will stain the skin a dark reddish brown color, which gradually fades over time.

Henna is also believed to have a cooling and soothing medicinal quality, and so it is often applied to broken and sprained bones, torn muscles, and dried, cracked hands and feet. Fadma explained that dry henna is often rubbed on the newborn's body to protect and strengthen the skin. She often recommended that women suffering from the intense summer heat of southeastern Morocco apply henna on their heads to keep themselves cool. Henna, she explained, not only strengthened the hair but also penetrated the body through the scalp to be absorbed by the bones, teeth, and eyes. She attributed her good health to the *baraka* that comes from a lifetime of using henna.

Herbs were only part of Fadma's healing technique. Her daughters explained to me that Fadma herself had *baraka*, ensuring that her cures healed her patients. Her *baraka* came from the twins she gave birth to thirty years ago. Women who give birth to twins are believed to be particularly fertile, and it is this connection to fertility that gives them *baraka*. In addition, the amber, roots, and herbs she crushed to make incense, and the antimony she ground to make kohl all had medicinal, healing qualities. Kohl, for example, not only decorated the eyes but also cleansed them of impurities. Amber, mixed with other herbs and roots, could be used to create incense. She made incense to cure mental illness and protect people from the evil eye.

The evil eye is connected with envy and covetousness. A person's good fortune, health, or looks can cause others to be jealous and activate the evil eye. The evil eye is greatly feared by people living in southeastern Morocco. Bad luck, illness, and even death are often attributed to the evil eye. The first glance of a person is considered the most dangerous. Hence, people wear different types of jewelry and pendants on the body to attract attention so that the envious gaze focuses on the attached object rather than the actual person. Fadma often made small leather and cloth pendants for children to wear, filled them with herbs, and decorated them with small beaded scorpions or cowrie shells to attract the negative first glance. Fadma also made beaded necklaces with a small copper or silver knife at one end. She explained that when an infant wears this necklace, the pointed knife pendant is protective. The blade of the knife can puncture and burst the evil eye and protect its wearer.

Fadma Lhacen treated the large number of women who visited the house every week with invariable generosity. Many a time I saw Fadma and her daughters stop their daily housework to welcome their female visitors with bowls of fresh cow's milk and plates of dates. After a woman had explained her problem and the appropriate procedure was performed, she often stayed for lunch or dinner and sometimes even spent the night. I asked Fadma how much she was paid for her services.

She explained that she never demanded a particular fee, but the women gave what they could.

Women would give Fadma an offering at the beginning of the healing session. This offering, which could be anything from a handful of dates to a couple of coins, was called the *ftouh* (opening), since it opened the way for Fadma's healing to begin. Often these offerings were also called *tisent n'ufous* (salt of the hand). The general idea behind these offerings was not to leave Fadma's hands empty, because emptiness symbolized the end of *baraka*. White substances, such as salt, are associated with purity and blessings and are believed to protect people from harm. For example, small children often wear bags of salt around their necks to protect them from the supernatural beings referred to as *jnoun*.

The *jnoun* (plural of *jinn*) mentioned in the Qur'an coexist with humans; however, their relationship with humans is somewhat precarious. The *jnoun* can be helpful but can also cause great harm. If the *jnoun* are provoked into action, they can cause serious illness or death and may even possess a person's body. It is believed that possession by the *jnoun* causes epilepsy, strokes, convulsions, and mental illness. A person whose body is possessed by the *jnoun* is said to be *mejnoun*.

In southeastern Morocco, people believe that the *jnoun* exist in entrances to villages, doorways, drains, and wells. Pouring hot water down a drain is dangerous, because the *jnoun* may be burnt and take revenge against the person. People are also instructed never to sit in thresholds, because these passageways should be clear for the *jnoun*. Passages can be physical and also temporal, and people moving through one of life's passages, such as circumcision, marriage, or motherhood, are especially vulnerable to attacks by the *jnoun*. For example, the period immediately after childbirth is precarious for the mother as well as the child.

Fadma did not heal illnesses caused by the *jnoun*, but she did offer advice for protection from them. She advised new mothers to never leave their children alone, and recommended that they constantly burn a candle for the first seven days of a child's life to discourage the *jnoun*, who dislike bright spaces and bright objects. White substances, such as salt, are also used in southeastern Morocco as protection against the *jnoun*. As already mentioned, a child may wear a small packet or amulet filled with salt and other herbs for protection. Salt guards these children from the *jnoun*, who are referred to in Tamazight as *wida tsantil tisent* (those who are hidden by salt).

Women in southeastern Morocco are particularly concerned with childbirth, since a married woman's status is directly related to fertility. A married woman increases her status in her husband's household, where she typically lives after marriage, after she has several children. Men often divorce infertile women or may take second wives. Most of Fadma's clients were women who wished to get pregnant. One story Fadma frequently recounted was how she cured a woman of infertility. This woman had tried without success for years to get pregnant and was frantic. Fadma administered certain remedies and gave her a package of herbs to mix with water and drink once she returned home. Two months later the woman was pregnant and returned with her husband to Fadma's house, bringing numerous cones of sugar and a sheep. Moroccans value the tall, elegant cones of sugar used to make mint tea; the whiteness of the sugar is associated with the moral qualities of purity and goodness.

Generous payments such as this were rare, however, and Fadma typically did not even look at the sum a woman gave her until after the woman left her house. I often asked her whether she really made any money, considering the food she gave her guests. Fadma explained that part of her *baraka* came from not making money from the healing she did. In fact, she had to do it simply because she could. It was her duty to share her *baraka*, and this made her a good Muslim.

Occasionally, a man appeared at her doorstep asking to be healed. Fadma always refused and recommended that he visit a male healer. She felt it was inappropriate for her to touch the body of an unrelated man; therefore, all her patients were women. This mentality and the paucity of men in Fadma's household meant that women could come and go freely and not worry about running into men. In southeastern Morocco, encounters between unrelated men and women were discouraged.

Tattoos, Dress, and Islam

The presence of so many women in the household gave me the opportunity to ask them about Berber art, which was the subject of my research. Until the 1950s, most members of the Ait Khabbash lived a nomadic lifestyle, traversing the desert with herds of camels, goats, and sheep. Like many nomadic groups, the Ait Khabbash created the kinds of art that were easily portable, such as textiles, and adorned their bodies with tattoos and jewelry. I often questioned women about the tattoos that adorned their faces and wrists. While they were happy to share their stories with me, most were ashamed of the markings on their bodies. Some women swore that they never wanted tattoos and blamed their tattoo marks on aggressive childhood friends, who they claim held them down and forcefully tattooed them. Fadma was more honest and explained that her mother's friend had tattooed her forehead and chin simply because everyone else was doing it. She explained that there were no specialists; friends or family members tattooed most of the girls at home. Women used materials that were readily available to create their tattoo marks. Soot from the bottom of a cooking pot traced the design on the skin. A knife, a needle, or palm tree spike pricked the design into the skin. Alfalfa was placed on the wound to give the design a green color.

Tattoos had no religious significance, but were public symbols of identity and marked women as members of the Ait Khabbash. Since Ait Khabbash cultural expression is typically associated with rural life, many women disliked their tattoos because they made them look *aroubia* (like a woman from the countryside). More important, Fadma explained that tattoos permanently altered the human body, God's perfect creation, and were therefore prohibited by Islam. In addition, tattoos are an obstacle that stopped water from reaching under the skin, creating a type of coating that renders ritual ablutions ineffective. Religious education and social pressure gradually caused Ait Khabbash women to abandon their practice of tattooing the foreheads, noses, cheeks, and chins of adolescent girls, and today it is rare to see a woman younger than forty with tattoo marks. Occasionally, women even used various chemicals to burn unwanted tattoo marks from their skin.

Fadma always tried to learn more about the Qur'an, and when her sons visited she frequently asked them to recite passages. Until recently, most Ait Khabbash women never attended school and did not learn how to read and write Arabic or French. Therefore, Berber women are often unable to read Islamic religious tracts or the Qur'an, which are written in Arabic. In Fadma's village, women do not pray in the mosque and do not attend Friday sermons. Although Islam does not prohibit women from praying in a mosque, local cultural practices discourage women from praying in public places. However, on religious holidays women gather together in a local house to pray.

Fadma explained that when she was a young woman, she knew only enough about Islam to recite her daily prayers. Her understanding of the Qur'an expanded greatly in the 1970s when the local *fqih* (religious scholar) decided to hold special lessons for women. Fadma gained knowledge of the Qur'an during these weekly meetings.

Special religious lessons are still held for illiterate women in Fadma's village, the only change now is that university-educated women lead them. Although Fadma no longer attends, she encourages her daughters to go. While Fadma holds fond memories of the classes she attended with the *fqih*, she told me that it was improper for an elderly woman like her to study with younger women. Some of these women have never been married, and Fadma explained that if the Qur'anic lesson dealt with sexual relationships or other sensitive matters, the younger students would feel embarrassed in front of their elders.

Although Fadma welcomed increased educational opportunities for women and even sent her youngest daughter to school, she resented other societal changes that had influenced her Berber group in recent years. In the past these Berber women controlled ceremonial life, and this gave them a certain spiritual power. Women gave life to the arts of Berber society and preserved many of their artistic and cultural traditions. Women had children, provided most of the childcare, protected the young, taught them about their Berber history and language, and, like Fadma, healed the sick through their knowledge of herbs. Fadma felt that elderly women no longer received the respect they once had in the society. She cited the example of her eldest son who moved out of her house with his wife and children. Although his job required that he live in another city, Fadma said that in the past a son always left his wife and children behind with his parents.

Societal transformations, such as increased educational opportunities, access to television, and the introduction of imported goods, resulted in changes in the artistic and cultural traditions of this Berber people. In addition, their transition from a nomadic to a sedentary lifestyle meant that many Berber traditions came under the scrutiny of their Arab neighbors. For example, Fadma explained that her youngest daughter prefers to wear the long black *lizar*, typical of Arab women, after some people in their village told her daughter that it was immodest for her to wear the brightly embroidered Berber *tahruyt*.

Southeastern Morocco is religiously conservative. Married and unmarried Arab women there often live secluded lifestyles. If an Arab woman leaves her house, she covers herself with the *lizar*, which completely conceals her body from head to foot. The *lizar* has a practical purpose and protects women from the hot sun and sandstorms characteristic of the area. It is also a modesty garment, and the large numbers of *shurfa* living in the region have influenced its conservative form. The *shurfa* are people who trace their ancestry to the Prophet Muhammad. Arab women who are *shurfa* pride themselves on their conservative behavior and strict style of dress, because they feel it reflects their closeness to God and their descent from the Prophet.

The conservative dress of the *shurfa* has influenced the use of a modesty covering, such as the *tahruyt*, by Berbers. Most women living in rural areas of Morocco, such as Fadma's Berber group, work in the fields, manage livestock, and gather firewood and water from great distances. While these women cover their hair with headscarves, they typically do not wear modesty coverings, because it hinders their ability to perform strenuous manual labor. Most Arab women in the Tafilalet oasis do not perform the heavy manual labor of Berber women, because their values call for a severe restriction of contact between unrelated men and women. Fadma recounted how her Arab neighbors often criticized her for leaving her house to collect food for her livestock, warning her that she might encounter unrelated men while working outdoors. Arab women, except women of very low status, never left their houses to collect firewood or work in their husband's agricultural plots.

Fadma resented these criticisms, stating that she needed to support her family. Fadma's husband died in the early 1980s after being ill for many years, and Fadma was responsible for the upkeep of her family. Regardless of whether her husband had been ill or not, I suspect, Fadma would have continued to work outside her home. Fadma told me that when her husband was alive, he often tried to help her collect firewood. Fadma would angrily refuse his help, telling him that he should not be seen doing "women's work," and send him on his way.

Fadma's group is one of the few Berber groups that do wear modesty coverings. However, the *tahruyt* worn by these women is made from deep blue indigo-dyed fabric. This veil is adorned with shiny silver sequins that shimmer in the intense sun, highlighting the vividly embroidered geometric and vegetal motifs. These striking garments contrast with the dry, brown landscape of southeastern Morocco, which lies on the fringe of the Sahara. However, their covering is more than a modesty garment, for its symbols and colors are similar to those found in other Berber textiles, symbolically connecting women to concepts of fertility. Indeed, it can be said that the covering worn by women of the Ait Khabbash, that is Fadma's Berber group, acts as a compromise between Islamic requirements of modesty and local forms of artistic expression.

Weaving and the Soul of Textiles

Woven textiles are a form of Berber artistic expression that is slowly declining over time. Fadma sighs when she tells me that her two youngest daughters refuse to weave. Much of the status of females once came from the ability to weave, but today many young women are not interested in textile manufacture. The importance of weaving for these Berber people is connected with their nomadic history. Berber women once wove tents for their families from goat and sheep wool. Women also used wool to make blankets, carpets, and clothing for their families. Weaving is a collective activity that requires many hands to set up the loom and finish the textile, creating a bond between women.

Wool and working with wool were so important to Berber societies that wool took on a spiritual significance for them. Wool is associated with the fertility of the land and God's generous nature, since it is God who supplies the rain that makes the grass grow and supplies the herds with food. Hence wool has *baraka*, and when people wear woolen clothing, some of the *baraka* of the wool is transferred to its wearer. Fadma told me that in the past women who did not know how to weave found it difficult to get married. Fadma believes that a woman who makes forty carpets during her lifetime receives the ultimate *baraka* or blessing: her passage to heaven is guaranteed when she dies.

While wool symbolizes the fertility of the land, the loom symbolizes the creative powers of women in many Berber societies. There is a connection between weaving, human creation, and the life cycle, and when the warp threads are attached to the loom, it is metaphorically born. The textile comes alive and is believed to have a "soul," or *ruh*. Textiles are equated with humans: they pass through the life cycle— and they also reflect the creative power of women. Fadma explained that textiles gradually age as they are woven and pass through youth, maturity, and old age. Just as the *jnoun* are drawn to people in vulnerable transition phases, such as birth, circumcision, and marriage, so too are they drawn to weavers, who must protect themselves from attacks by the *jnoun*. Prior to weaving, a woman sprinkles the area surrounding a loom with salt in order to remove the *jnoun*.

Women give life to a textile and they also take its life away. When a textile is finished, the warp threads are cut and the textile dies. Fadma once told me to watch as she cut a blanket she was weaving from the loom. She splashed the blanket with water and said, "Drink, loom. You will drink tomorrow in heaven." Just as Muslims wash a dead person before he or she is buried, the textile is splashed with water because it is considered to be dead.

Postscript

When I left Morocco after two years of my study and returned to the United States, I realized just how much I had learned from Fadma

Lhacen. I return to Morocco and visit Fadma practically every summer. Fadma no longer wears her brightly embroidered Berber *tahruyt*, claiming she must dress more modestly now that she is getting older. Instead, she wraps a long piece of black cloth around her and heads out into the hot sun to collect dates for her animals. Although her three sons send her money each month and plead with her to stop working so hard, Fadma Lhacen insists that it is her duty as a good Muslim to continue helping her family as long as she can.

Fadma continues to maintain her reputation as a healer, and if Fadma is tired or not feeling well, her oldest daughter treats the visiting women's illnesses. Gradually her eldest, divorced daughter has begun to earn her own reputation as a healer and a carrier of *baraka*. Many men have asked for her daughter's hand in marriage. Fadma neither discourages nor encourages her daughter to accept their marriage offers. Rather, she always tells her daughter to make her own decisions, and her daughter simply tells the men that she prefers to take care of her mother as she advances in age rather than get married again. Fadma plays an important public role in her local community as a healer, sharing her knowledge of herbs and her gift of *baraka*. As an older Berber woman of the Ait Khabbash, she is an important resource for other women. But Fadma Lhacen's own goal is to live her life as a good Muslim, supporting her family and others in her community.

CHAPTER 3

Master Singer of Pakistan: Baba Shafqat

(Pakistani Religious Musician)

John and Linda Walbridge

There are still people in Baba Shafqat's family who are sorry he didn't become a movie star. He is an old man now, but when he was young, he was handsome, and he still has a splendid voice. The Indian movie industry began as early as Hollywood, and itinerant theater troops had traveled among Indian cities long before that. Young Indians and Pakistanis dream of the glamorous life of an actor. One of Baba Shafqat's cousins had become a famous screenwriter in the fledgling Pakistani film industry. When he came back to their hometown of Gujrat—secretly, since his father had kicked him out of the house for the disgrace his career had brought on the family—Shafqat and the other boys of the household would hang on his wild tales of the movies and act out his scripts among themselves. Right from his school days, Shafqat acted in stage plays and began getting invitations to act in the movies and perform on the radio. There is still a publicity photo in his house showing a handsome young man in the self-conscious attitude of a 1940s movie star. In the end he bowed to family pressure and gave up his budding acting career to become a religious leader, like so many of his ancestors. He is still a performer though, and leads a group of young men who sing in praise of the martyrs of Shi'ism. It took a while for us to figure out just what else he was, for he crossed too many lines to be simply explained.

Figure 2 Baba Shafqat with His Young Men

We first met Baba Shafqat in Lahore, Pakistan, through his nephew Nadeem, a suave, roving diplomat for a Shi'ite foundation in London. Nadeem knew that Linda was interested in Shi'ism and thought that his uncle would interest her and be able to introduce her to other Pakistani Shi'ites. Baba Shafqat and Linda took to each other instantly. He was then in his mid-sixties, but he was wiry and energetic, leaning forward in his chair in our living room to talk excitedly about plans and the people Linda should meet. He wore traditional Pakistani dress and seemed far less sophisticated than his nephew. He had the beard of a pious Muslim, but it was neatly trimmed, as befitted a former future movie star. 'Ashura, the Shi'ite season of mourning and Baba Shafqat's busy season, was coming up, and he wanted us to see him and his group perform. Their job was to inspire people and make them weep for the wrongs done to their leaders thirteen centuries ago.

Grieving for Husayn

There is an old joke that a visitor to India noticed a procession of people wailing and beating themselves. He asked his guide what was happening.

"Oh," replied the guide, "those are the Shi'ites. They are mourning for their leader who was killed with all his followers. It is very sad."
"I am sorry to hear it. When did it happen?" said the visitor, who hadn't seen anything about it in the newspaper.
"A thousand years ago, I think," said the guide.
"And they have only just heard?!"

In truth, the grief of the Shi'ites does not fade with time, because it is the core of their religion and the mirror of their own sufferings. The Shi'ites were those who believed that the descendants of the Prophet Muhammad should be the leaders of Islam—"Imams" as they call them—but it was the Prophet's friend Abu Bakr, not his blood heir 'Ali, who became the first leader of the community. When 'Ali did finally become the leader twenty-five years later, it was only to face rebellion and an assassin's knife in the Iraqi city of Kufa. In 680, the Imam Husayn, the grandson of the Prophet and 'Ali's son, accepted a plea from his followers in Kufa to lead a revolt against the impious caliph Yazid. But Husayn was betrayed. Spies betrayed the plotters to Yazid, whose governor put Kufa under martial law. Husayn and his little party of supporters and relatives were trapped at a place called Karbala, a few miles from the Euphrates. The Shi'ites of Kufa never came to their aid. Tortured by thirst and facing overwhelming odds for ten days, Husayn's men were killed one by one. Last of all Imam Husayn died, facing his tormentors and fighting alone. His head and his women were taken as trophies to Yazid in Damascus.

The modern Shi'ites are the spiritual descendants of those Kufans who failed to come to Husayn's aid. Almost always a minority in the countries where they live, the Shi'ites have mourned Imam Husayn in grief and in guilt, remembering the wrongs done to their leaders and to themselves, remembering also their own failures and hoping for the day when the last of the Imams would return with the sword to rally his faithful followers and rid the world of injustice. It is a powerful, emotional mixture, giving Shi'ism a spiritual flavor very different from the rest of Islam, perhaps similar to what Christianity might have been if its formative experience had not been the resurrection but Peter's threefold denial of Christ. The memory of those tragic days at Karbala can be used for many purposes—to counsel patience in adversity or to appeal for revolution in the name of justice—but for all Shi'ites the most important time of the religious year is 'Ashura, the ten days when they gather to remember the sufferings and death of Imam Husayn at Karbala. It is Baba Shafqat's job to shape the emotions of those days and to give them meaning.

'Ashura—literally, "the ten days," the first ten days of the month of Muharram—is observed in various ways across the Shi'ite world. In Iran there are passion plays in which amateur actors reenact the sufferings and deaths of Husayn and his companions. Emotions can run high; actors playing the chief villains are occasionally lynched by the audience. In Pakistan there are processions in which groups of Shi'ite men compete to demonstrate their fervor. Linda and I went to one in a place called Model Town, a newer neighborhood of Lahore. The atmosphere was a curious combination of religious festival, celebrity funeral, and Fourth of July parade. Vendors sold snacks, religious artifacts, and whips. Water trucks, with the names of the sponsors painted on their sides, distributed drinks to passersby, as is fitting on this day honoring those who had endured thirst. Groups of young men, overseen by sweating and overweight middle-aged men, competed for attention. Some led elaborately decorated horses and carried black banners, both commemorating Imam Husayn, the fighting hero. Most, though, stripped to the waist and beat themselves, either with their own hands or with whips tipped with knives, striking themselves in time to chants in honor of the Imam. There was a good deal of blood. Groups of women in black veils watched. The young men preened themselves, keenly aware that not all the veiled women were old. The groups of men converged on a central point from several roads. A helpful young man directed us to where the various processions would meet in a climax of self-inflicted pain and religious exultation. "It is just awful," he said. "You'll love it!" We met a Pakistani emigrant to Scotland, his clothes stained with blood, who came back every year with his fascinated children.

Baba Shafqat, however, participates in a more decorous observance of 'Ashura, the *majlis* (meeting). These are held either in an *imambargah*, a kind of Shi'ite mosque reserved for such purposes, or in the courtyards of the homes of wealthy individuals. In Pakistan these events typically last about an hour and a half. There are several "acts." Qur'an reciters, imams reciting prayers, perhaps schoolchildren chanting memorized poems—leading up to one or two headliners—a well-known preacher, a famous reciter of poetry, or both. Some majalis feature famous performers. Baba Shafqat and his young men have a wide reputation, much like the reputation of good gospel groups in the United States.

On the twelfth day of Muharram, Baba Shafqat came to take us to an *imambargah* in Moghulpura, a section of Lahore. In the car were his driver and a Lahore policeman, in place of his usual armed family retainer. This *imambargah* had been destroyed in 1986 by Sunni terrorists, and there had been several recent bloody attacks on Shi'ite religious gatherings. But Baba Shafqat assured us that the *malang* (dervish) of the *imambargah*, Sayyid Khwaj Muhammad, was a very brave man and we would be quite safe. The area was crawling with police, and the *imambargah* had its own armed guards, but that night it was peaceful. We sat for a while in the courtyard with the *malang* and his disciples. A *malang*, Baba Shafqat told us, is a hippie. He meant that Khwaj Muhammad was a sort of wandering monk whose spiritual knowledge came from experience, not from books. Since childhood, Sayyid Khwaj had been the disciple of a village holy man whose knowledge came directly from the Shi'a Imams. Most likely, both teacher and disciple were illiterate, or nearly so. When his master died, Sayyid Khwaj had walked to Lahore and built his own *imambargah*. Baba Shafqat thought very highly of him. When people started to arrive, Linda went off with the ladies and the rest of us went to the office for tea. The *majlis* began without us. Baba Shafqat conferred intently with his young men over a binder full of poetry and notes. After a while we went out and listened to the first speaker, a screaming preacher with an earsplitting sound system. After another quarter hour Baba Shafqat and his young assistants took over. Baba Shafqat spoke for a few minutes about the sufferings of the Imams, then he and the young men burst into song. It was unexpected and very beautiful, sounding much like gospel music. He alternated between speaking and singing, working the emotions of the audience to a crescendo. He spoke of the Imam Husayn on the last day of his life, tortured by thirst, his infant son dead in his arms from a stray arrow. People began to cry. He sang the last poem through his own sobs, then he skillfully brought down the emotional level. The congregation dispersed, emotionally refreshed in the curious way it happens at such events. We went back to the office and had dinner, eating our chicken with our fingers and sitting cross-legged in a circle around a tablecloth on the floor, in the company of Baba Shafqat, his young men, Sayyid Khwaj, and the Moghulpura magistrate, a Sunni who was there to supervise security arrangements.

A Very Old Family

It wasn't entirely clear to us that Baba Shafqat's elders, having forbidden him to become an actor, would have been any happier about his career as a sort of singing preacher. Pakistan is a fragment of the once united British India and still has some features of Indian culture, including the remnants of a caste system. Baba Shafqat was born into the highest of Muslim castes, for he is a sayyid, a descendent of the Prophet and of the Imam Husayn himself.. The family has a scroll documenting their descent from the Prophet. It was first written centuries ago, and the later generations have pasted on additional pages. His family has lived in the little town of Gujrat for four hundred years, intermarrying so often that there are rueful jokes about inbreeding in relation to some of the odder cousins. In short, as Baba Shafqat's nephew observes, they are a family of poor aristocrats continually lamenting their former glory.

Four hundred years ago India was rich and at peace, under the rule of the Muslim Mogul dynasty. Ambitious and talented Muslim religious scholars migrated to India in search of opportunities and employment. Among them were the ancestors of Baba Shafqat. As befitted members of a holy family, their descendents seem to have mostly been religious scholars of local distinction. One could probably reconstruct the family's history with the aid of the scroll, but it would tell us little more than that the family went on from generation to generation, marrying cousins or other sayyids, occasionally producing someone of distinction, but more often producing modest, dignified men who preserved the prestige of their ancestors and who taught and practiced their faith as their talents and inclinations allowed. For much of that time, they were among the dominant families of Gujrat and its district, living a life of reasonable wealth and impeccable respectability. There were occasional setbacks. One of Baba Shafqat's grandfathers fell in love with a non-sayyid servant girl and squandered much of the family's property in pursuit of her. The family still remembers this, because the descendents of servants now own land that once belonged to them and are now wealthier than their former masters.

Despite his grandfather's indiscretion, Baba Shafqat grew up secure and spoiled in a household that was still wealthy and traditionally religious, living in a house that reflected the courtly traditions of the Moguls. Even today, Baba Shafqat's Urdu is so full of antique literary words and expressions that it can be nearly incomprehensible to the younger members of the family. He was a spoiled child; an only child who was raised with his mother's much younger sister whom his father had adopted as his own daughter. In those days the family was dominated by Baba Shafqat's father and his father's cousin.

The cousin, Sayyid Imdad, was a turbaned traditional religious scholar, a man of scholarship and learning in Shi'ite laws. He had regular visitors from other parts of the world and was a great friend of Grand Ayatollah Shari'atmadari in Iran. The Maharaja of Poonch in Indian Kashmir—one of the native princes of the old British India—was an admirer and invited him to lecture in his little kingdom. The family remembers him best for the library he built up. There is no one left in the family who can read most of the books, for they are in Persian and Arabic and many of them are handwritten manuscripts. However, the books are all brought out into the courtyard and dried in the sun every spring so they won't be spoiled by the dampness. Sayyid Imdad was part of the old tradition of religious learning that the family still respects but no longer emulates.

Baba Shafqat's father, Ahmad 'Ali, was something else altogether. He was a *pir*, a mystical leader with a popular following, and was more like Baba Shafqat's *malang* friend, Sayyid Khwaj. Every year he led pilgrimages to the holy places of Shi'ism in Iran, Iraq, and Syria (where the severed head of the Imam Husayn was finally laid to rest). Unlike the learned and intimidating Sayyid Imdad, Ahmad 'Ali was approachable and had a wide following among both the elites and ordinary people. He taught Persian, which was still the language of gentlemen in conservative Indian families in his days. In particular, he had a reputation for being a simple saint. He personally adopted children in India and Iran and secretly paid for the education of many servants and otherwise indigent or mistreated folk. He would leave the house in the morning and return barefooted and shirtless, having given away his belongings to people who needed them more than he. As a result, his personal reputation was immense, and people came from far and wide to sit with him, hear his words of wisdom, and benefit from his spiritual aura. He was also occasionally a victim of charlatans and con men and lost a great deal because he trusted people. He was responsible for the conversion of many animists, Hindus, and Christians to Shi'a Islam, including the people in the nearby village of Bashna, who still pay him and his descendants particular reverence.

When Shafqat became a young man, he faced difficult choices. In 1947, when he was fourteen, British India was partitioned into two independent states, the Republic of India and Pakistan. Gujrat was in Pakistan, and young Pakistanis were excited about the prospects of building a new and modern Muslim country. For the most part, young Pakistanis had little interest in the old religious scholarship and customs, and even the languages were changing. Young Pakistanis of Shafqat's generation wanted to learn English and Urdu, the languages of the new Pakistan, not the Persian and Arabic of the old Indo-Islamic aristocracy. And, of course, Shafqat was stagestruck. In the end, he joined the Pakistani Air Force with his cousin Sajjad.

That did not work out, and Shafqat's departure from the Air Force was not amicable. At some point, there was a showdown with his father. The family does not talk about what exactly happened, but Shafqat gave up his acting ambitions and Sajjad immigrated to England. Shafqat assumed his father's role as a local spiritual leader. Thus, while his cousins and their children took up modern careers in Pakistan, the Persian Gulf, England, or America, Baba Shafqat stayed behind in Gujrat to carry on the family's traditional religious responsibilities. With his father's death in 1974, he became the leader of the family, exercising a kindly despotism over cousins, nephews, and nieces scattered across the world. For example, he had his own daughter marry the son of his cousin Sajjad in England and assigned one of his disciples from Bashna, who had moved to Long Island, to telephone the couple every day and check on their well-being.

The *Pir* of Bashna and the Sayyid of the Mahalla

Baba Shafqat's return to the traditional religious life was not so hard for him as one might think, because, as we have seen, he remained a singer and a performer. He brought an incongruous echo of the movie star, which he could have been, to his laments for the martyred saints of Shi'ism. Moreover, he is deeply religious in the tolerant way of the old Muslim mystics of traditional India. The Indian Subcontinent was and still is a country full of religions. While there is no lack of sectarianism, too often expressed in gruesome acts of intercommunal butchery, local people often respect the saints and shrines of their neighbors, especially if a holy man is obviously holy. Thus it was that the *pirs*, the Muslim mystical leaders, were typically the pioneers of Islam in the towns and villages of India.

"Pir" is a Persian word that simply means "old man." The *pir* is the spiritual master to his disciples and a guide, mediator, and a source of blessings, advice, and practical help to those who are linked to him less formally. Since Baba Shafqat's father was a *pir*, his father's admirers expected to him to fill his father's shoes. We saw Baba Shafqat in his role as the master at the *imambargah*. The relationship between master and disciple is deeply engrained in Islamic culture: teacher and student, craftsman and apprentice, mystical guide and seeker, father and son. The young men who sang with Baba Shafqat were his disciples. They were obviously relaxed with him and loved him. When he needed something, they took care of it, usually without being asked. When Baba Shafqat spoke, they were silent. They spoke to him in low voices. When he gave instructions, they obeyed without question. In a few decades they will exercise the same comfortable and unquestioned authority over their own disciples. Such chains of masters and disciples stretch back through the centuries in many areas of Islamic life. There is always an element of religion in the relationship; for example, while the master may be a craftsman and the disciple his apprentice, they may trace back the origin of their craft to a prophet, and the guild of craftsmen and apprentices may have old and deep relations with one of the local shrines or mystical orders.

Baba Shafqat exercised a less direct but equally strong authority in the village of Bashna. He took Linda there one day to meet his followers. Many of the villagers had been converted to Shi'ism by Baba Shafqat's father. Bashna is about twelve miles from Gujrat, though it takes forty-five minutes to drive there because of the bad roads. It is an agricultural village of about eight hundred houses, with a mix of older mud brick and newer concrete houses; the latter usually belong to people who work in the city or abroad. The village was only recently electrified. Education is still a problem. There is a government school in the village, but it has room for only about a hundred pupils. There is a *dini madrasa* (an Islamic religious school) that serves more students, but only boys, and its standards are even lower than that of the government school. There were black flags above many of the houses indicating that the families were Shi'ite. Other houses flew the flags of one of the Sunni religious parties. Baba Shafqat took Linda to one of the better houses, where lunch was served, after which they visited several other Shi'ite homes. Most had pictures of Baba Shafqat or his father. People came to him to be blessed, and he called them all "brother" or "sister." He does not function as a cleric, for he is not a teacher of religious knowledge. The villagers see him as a man whose spiritual power can be passed on to them to protect them. He does not have the same relationship with these people as he does with his disciples. He certainly does not exercise the same day-to-day authority over them. They obviously revere him, and he probably receives the Shi'ite religious tax owed to sayyids from them. If there are problems—difficulties with the government, an expensive medical treatment, a shortage of money for school fees—they might come to him.

Baba Shafqat is also a leader in Gujrat, especially in his own neighborhood, which has both Shi'a and Sunni families. To start with, he is a leading figure among the local Shi'ites. That afternoon, he took Linda to the *imambargah* for a service to commemorate the death of his father and the first anniversary of the founding of an organization to support his own work. They passed through the cemetery, where a shrine marks the grave of the first member of the family who came to Gujrat four centuries ago. The service, mostly speeches by neighborhood dignitaries, had been going on for some time when they arrived. Baba Shafqat's young men sang. A Sunni preacher spoke about Karbala and the Imam Husayn "from his perspective." As is the custom at Pakistani religious events, loudspeakers broadcast the proceedings to the neighborhood. Eventually, the people moved out through the cemetery and formed a procession, led by a band of bare-chested young men beating themselves rhythmically with their fists and by the elaborately caparisoned horse that represents the Imam Husayn. The procession wound through the neighborhood to Baba Shafqat's house, where everyone was served lunch. Although he doesn't receive the same spiritual veneration in his own neighborhood, he is nevertheless respected and plays the same role as mediator when there are problems.

The Narrow Road

On his business card Baba Shafqat gives his title as "Faqir," dervish or, roughly speaking, "monk." The card also lists several national and provincial Shi'ite organizations of which he is an officer. The contradictions are a reflection of his life, for despite his quite genuine air of simple spirituality, he must walk a narrow political road among many powerful contending forces: his own people, the Shi'ite communities and organizations in Pakistan, the powerful centers of Shi'ite authority in Iran, who are anxious to win influence among the Pakistani Shi'ites through their money and prestige, and the increasingly vicious Sunni militant groups and their Shi'ite counterparts.

Relations between the Shi'ites and Sunnis in Baba Shafqat's own neighborhood are correct but uneasy. Baba Shafqat himself is respected by all. Linda actually stayed with his Sunni next-door neighbor. Effectively, the two families live like a single household. Each house surrounds a courtyard, but there is a roofed area in the neighbor's courtyard, so the cooking is done there when it rains. Baba Shafqat's wife teaches Qur'an to a number of small children of the neighborhood. Early each morning they appear in the courtyard with their slates. His wife is respected for this, for it is a good thing to teach, especially to teach the Qur'an. However, there are traces of tension. The Sunni neighbor commented that Shi'ites did not marry Sunnis, and a few people had left when the Sunni preacher came on stage at the *imambargah.* Likewise, Bashna has divided allegiances. There had been violence there between Sunnis and Shi'ites, but Baba Shafqat had worked to bring peace.

Baba Shafqat had good reason to bring along an armed policeman when he took us to the *imambargah* in Lahore. Many Shi'ite leaders were assassinated in recent years. At one time, for example, a Sunni militant group was systematically killing Shi'ite doctors in Karachi, Pakistan's largest city and a cauldron of ethnic and religious tensions. The trouble went back to the 1977 coup that brought General Zia-ul-Haq to power. Having overthrown and hanged an elected leader, Zia was short on legitimacy. He was, however, a pious Sunni Muslim and an admirer of the most influential Islamist group in Pakistan, the Jama'at-e-Islami. He thus began a headlong program of Islamization of law and society in Pakistan, thus gaining Islamic, if not democratic, credentials. Soon after, the United States began supporting the Afghan Islamic groups who were fighting a jihad, or holy war, against the Russian occupation of their country. Much of the American aid to the Afghan groups was funneled through Pakistan. Saudi Arabia also contributed to the cause in various ways, in the process supporting its own variety of very conservative Islam. This gave a tremendous boost to the most militant of the Islamic groups in Pakistan, who received money, weapons, and training from America, Saudi Arabia, and the Pakistani government. Moreover, a provision of Zia's Islamization program allowed these groups to open religious schools, madrasas, with government funds. The effect was to create a number of well-armed, highly militant groups with large networks of schools spreading their views.

The Shi'ites and other minorities watched these developments with alarm, as did more liberal and secular Pakistanis. The Islamic law imposed by Zia was Sunni law of a quite conservative variety. Shi'ites became indignant on finding that they were expected to pay a religious tax to the government that would support Sunni religious causes and that was contrary to Shi'ite law. The government backed down in that case, but the new Saudi-influenced militant groups increased their influence nonetheless—getting their version of Islam written into the government religion textbooks, for example. With the Soviet withdrawal from Afghanistan, the militant groups turned their attention to establishing their own form of Islam in Pakistan. Although relations between various religious groups in Pakistan had traditionally been fairly good, the militant groups tended to consider the Shi'ites as infidels. There were Shi'ite groups who fought back, supported by the revolutionary Shi'ite government of neighboring Iran. The result was a series of bloody tit-for-tat terrorist attacks, including attacks on Iranian diplomatic and cultural facilities and staff and on both Shi'ite and Sunni mosques and religious sites. The authorities were reluctant to intervene, since police and judges had also been attacked by the militants, and the militants also enjoyed the support of a part of the Pakistani government.

Baba Shafqat himself is a tolerant man, devoted to his religion but respectful of the views of others. He is proud of his good relations with his Sunni neighbors. He was delighted with a gift I gave him, a children's book of scenes from the Bible, painted by an Indian nun in traditional Indian style. He believes that each religious community should cultivate its own identity but be friendly with others as brothers in faith. However, the efforts of the militants to enforce a conservative Sunni orthodoxy on Pakistani society threaten his community and his own work. Shi'ites are not Muslims, the extremists insist. Certainly, the extremists would not tolerate the sort of musical veneration of the Shi'ite Imams that is Baba Shafqat's lifework; music, the militant Sunnis believe, is forbidden by Islam, as is the veneration of saints. And, of course, it is perfectly possible that the gunmen will target him one day, so he travels with an armed guard, one of his followers from Bashna. Baba Shafqat is also a senior official of the provincial branch of one of the Shi'ite organizations set up to counteract the militant Sunni groups.

He walks an equally narrow road with Iran. Revolutionary Shi'ite Iran is the protector of the Shi'ites of Pakistan, just as Saudi Arabia is the protector of the conservative Sunni groups. Iranian money supports the activities of the Shi'ite organizations of Pakistan. A foundation supported by Ayatollah Khamane'i, the religious leader of Iran, had supplied the funds to start Baba Shafqat's Pakistan Association of Reciters, which would train Pakistani Shi'ites to perform Baba Shafqat's kind of musical recitation. The same foundation has also founded Shi'ite schools in Pakistan. Baba Shafqat has visited Iran a number of times and has strong connections there. A picture in his home shows him being received as an honored guest in Qom, the Shi'ite religious capital of Iran.

At the same time, he is cautious about these connections. He is a patriotic Pakistani, a veteran of the air force, and he has no desire to see the Pakistani Shi'ites dominated by Iranians. Moreover, there is much about the religious government of Iran that he would not approve of. Shari'atmadari, the Grand Ayatollah who was a close friend of his father's cousin Imdad, was ruthlessly persecuted by Khomeini's supporters after the revolution of 1979. On the recommendation of a prominent Pakistani Shi'ite cleric, Baba Shafqat follows Ayatollah Sistani in Iraq in matters of religious law, and not one of the Ayatollahs in Iran. There is also some doubt about whether the more conservative Iranian religious authorities would approve of his musical activities any more than the Sunni militants would. Thus, while he receives the overtures of Shi'ite leaders in Iran hospitably, he keeps his distance.

Though Baba Shafqat has chosen to play the traditional religious role of his ancestors, his early experience as an actor still stands him in good stead. Had he lived two centuries ago, his life would have been simpler. He would have carried out his task of reminding his people of the sufferings of their leaders of long ago, and no one would have questioned him. His own community would have judged him by his piety, and his neighbors from other religions would have judged him by his holiness. Things are now more difficult as the world changes around him. His message retains its significance, for the story of those who suffered for demanding justice from their rulers is, if anything, more relevant now than it was when the first member of his family came to Gujrat. But now there are new and intolerant voices in his world—Sunni militants who are convinced that there is only one true form of Islam, Shi'ite leaders in Iran seeking authority over the Shi'ites everywhere else, and always the insistent distractions of the modern world with its indifference to the old forms of religion. These forces are too powerful for him to oppose directly, so he must negotiate his way among them.

Still, Baba Shafqat is a happy man. He emphasizes tolerance and understanding as supreme values and is opposed to extremism in all its forms, but he is especially opposed to religious extremism. Religion and faith, he insists, are about love and engagement, not resentment. Loving one another as human beings is the answer, and we should promote goodness and positive human relations regardless of caste, color, creed, or background. You cannot hate or be of bad character if you want people to love you. People only move close to good people. He has the good natured casual courage of Pakistanis with a cause. He believes that the message of Imam Husayn and his followers is still relevant to his fellow-Shi'ites and, if they will listen, to others. He thinks that his work makes Pakistan a little more peaceful than it might otherwise be. He loves his music and the young men who follow him and sing with him. And, as when he was a young man, he still loves to perform.

CHAPTER 4

Esmat Khanum and a Life of Travail: "You Yourself Help Me, God" *(Iranian Village Woman)*

Mary Elaine Hegland

In the village of ʿAliabad, back when Esmat was young, husband and wife did not interact much at all in front of other people, much less did they show affection for each other. They largely spent their lives in different social circles. During the day the husband went off to the fields, to a shop, or to a job in the city, while the wife stayed close to home, washing clothes or preparing food in her courtyard, and chatting with female neighbors and relatives. During leisure hours, the men might go to the mosque to pray, or sit near the gateway to the village watching people come and go, while the women visited family members and relatives in other sections of the village. On Thursday afternoons, women walked to the local cemetery, usually in the company of female relatives, to weep and pray for the souls of the dead. Even when attending a wedding or funeral, hosts would arrange one courtyard or home for men guests and another for women, so even there husbands and wives sat separately.

Women concentrated their emotional lives and social interaction on female family members, relatives, and neighbors; marital relations tended to be contentious or neutral. But even for those couples who got along well together, they did not focus on each other in front of others or show affection. Especially in front of older relatives, attentiveness to each other would have been considered rude and disrespectful.

Esmat Khanum and her husband were exceptions. They were sweethearts, other women told me, like the traditional lovers Leila and Majnun, or Shirin and Farhad. We just liked each other a lot, explained Esmat. He was her father's brother's son and they married when she was fourteen. For the first six months of their marriage, he worked in the city of Shiraz and came home one day a week, and she stayed with his parents in the village. Then she was taken to the city. Her husband rented a room in the courtyard where his brothers and other neighbors also lived. He worked as the night doorman in a hotel, so it was difficult for a young bride. She had to ask his brother's wife to come and stay with her at night, and that made her sister-in-law feel that she could interfere in their lives.

Going to the Shrine in Time of Trouble

One day Esmat and her husband were standing and talking, and in fun she put her hand on the back of his neck. On seeing this her sister-in-law turned to her brother and said, "That means that you are not a man, that you are *bighayrat* (without honor)." He hit Esmat, saying "Why did you do that in front of people?"

Esmat left the house and went to Shah Cheragh, the tomb of the son of an Imam, and sat there for two or three hours. She wanted help. Then she circled the tomb praying, "Son of Musa Ibn-e Ja'far, you yourself help me, so that others will not be able to interfere and ruin my life." At that time, men and women were not separated at the shrine. After a while, she noticed that her husband had come and was looking for her. He said, "Please forgive me for making you unhappy. I made a mistake. I shouldn't have said that to you. I know you didn't mean that when you put your hand on my shoulder." Esmat got up and went home with him.

After this, her husband left his job at the hotel and went to work at the Department of Post Office and Communications so he wouldn't be leaving her alone at night. Esmat no longer had to ask her sister-in-law for assistance, to come and stay with her at night while her husband was gone. Her sister-in-law could no longer feel she had the right to interfere in their lives. Further, Esmat's going to Shah Cheragh at a time of crisis, instead of to a relative, had avoided the problems connected with seeking help from yet other people. She had sought help from God, the Imams, and the imamzadehs, or sons of Imams.

Meeting Esmat, Her Early Widowhood, and Life in the Village

In the late 1970s, I went to the southwestern Iranian village of 'Aliabad to conduct field research. In the beginning of my sixteen months in the village, I wore a scarf and tunic over loose pants, but as the only woman in the village who did not wear a chador, I felt too conspicuous. My landlady suggested that I ask Esmat, a neighbor who was a seamstress, to make me a chador and do other sewing for me. I soon became a frequent visitor to the small courtyard and room where Esmat lived as a young widow with two sons and her mother. Her brother and his wife and baby lived in the adjoining room.

Although Esmat was basically illiterate, she was very intelligent and thirsty for knowledge. Her brother worked as a tailor in the nearby city of Shiraz and had joined a Qur'an study group. She learned much about Islam and current events from him. Esmat's mother took the bus into Shiraz to buy cloth. Her mother had also become a widow while she still had young children to care for. Even before her husband died, she had needed to bring in an income; her husband had worked as a trader in Beyza, a neighboring area, and didn't bring back much money for his family. Women came to their home to buy cloth from her mother, which Esmat then sewed into children's clothing, baby layettes, and the loose pants that everyone wore at home. She also made the traditional clothing of full pants, gathered long skirts, and long tunic dresses.

Women from all over the village came with their sewing errands to Esmat's home. While they looked over the cloth and told Esmat what clothing they wanted, they conversed. Esmat served small glasses of tea and offered a water pipe, and she kept up a stream of questions and comments. The women shared their information and perspectives about happenings in the village. Because of her central position in the village communication system and her abilities in soliciting and reporting information, Esmat was especially helpful to me as a foreigner in learning about events and relationships in 'Aliabad.

As for Esmat herself, she enjoyed a reputation as a highly religious, pious, and modest woman. Although she was a widow and had no man who was directly in charge of her, there was never a breath of gossip or critical comment about her. Others respected her because of her extremely circumspect behavior, her obvious devotion to God and the Shi'a holy ones, and the way she fulfilled people's expectations of a good Muslim woman. I remember thinking that her identity as a righteous and dedicated Muslim woman took the place of a husband in protecting her from any potentially insulting behavior and slander.

Esmat frequently uttered religious phrases and brief prayers. Very often she called upon God, the Imams, and the saints to help her. Before any trip, no matter how short, or any task, she would always first invoke the name of God, saying, *Bismillah al-Rahman, al-Rahim* (In the Name of God, the Compassionate, the Merciful). Of course, she faithfully performed the required five sets of prayers a day, telescoped into three periods for Shi'a Muslims. During Ramadan, the Muslim month of fasting, she woke early before daylight to eat a meal with her family. During the daylight hours of Ramadan, she abstained from eating, drinking, and smoking, and then enjoyed a special meal to break the fast in the evening after sunset, either with her own family or as an invited guest with neighbors or relatives.

As an illiterate, rural young female with very little income or property at her disposal, she was not able to engage in many religious activities that wealthier males and even urban females could practice. She could not read the Qur'an or other religious texts, or go on hajj (the pilgrimage to Mecca), or participate in Qur'an reading, reciting, or discussion groups. As a young woman, a widow, Esmat could not go to the mosque for daily prayer or for the Friday noon prayer. Rather, Esmat learned about Islam through oral means, through listening to her mother and to her brother. Further, by attending gatherings commemorating the martyrdom of the saints of Karbala and the suffering of their female relatives, she learned the stories and their meanings.

Esmat hosted a weekly *rozeh* (recitation of a martyr's story) in her home. I frequently sat in on these sessions. The blind *rozehkhun* (martyrdom story reciter) came to Esmat's home as part of his regular schedule. He chanted a story, perhaps a poignant account of the suffering of Imam Husayn's young daughters at the Karbala camp where Imam Husayn and his small force of seventy-two men met with defeat and martyrdom at the hands of the caliph's army in 680 AD. When nearing the climax of the story, Sayyid Habib's voice would grow hoarse with emotion. The women would pull their veils over their faces and sob at the thought of three-year-old Roqayeh crying in grief when presented with her father's decapitated head. After the recitation, in which neighborhood women joined, perhaps also bringing a few coins to make their own requests of the saints, Esmat served small glasses of tea and prepared a qalyun (water pipe). The visitors exchanged news before going back to their own worldly tasks.

During Muharram, the month of mourning for Imam Husayn and his band, wealthy villagers hired storytellers, and some years back, even players from the city to put on a series of theater nights in which they acted out the passion of the Karbala saints. Gifted with a fine memory, Esmat had become knowledgeable about Shi'a Islam from these various sources. She, more than anyone one else, shared religious stories and knowledge with me during my stay in 'Aliabad.

Esmat faithfully attended funeral gatherings for other villagers on the day of their death, the third day after death, the seventh day, the fortieth day, and the one year anniversary; during these gatherings she prayed for the soul of the departed and consoled the bereaved. On Thursday afternoons, she walked to the cemetery to pray for the souls of her own departed family members and relatives. She also visited the other groups clustered around the graves. She distributed dates and other fruits and ate food that others offered in their bereavement. Mourning ceremonies required no invitation, and people generally expected their neighbors to attend when they hosted meals to honor the saints. News of an upcoming event passed around by word of mouth. Because of her piety and wide circle of relatives and acquaintances, many of them on account of her sewing activities, Esmat was often invited to meals in honor of the martyrs and saints held in other sections of the village.

As much as possible, Esmat visited the shrines of the Imams and the imamzadehs to pray for their intercession and help with her family and worldly issues. Two small and rather neglected tombs, supposed to be those of descendants of the Imams, lay within the village environs, but for a fuller pilgrimage experience and hopefully more effective intervention, women had to travel outside the village. During the early period in Esmat's marriage when she and her husband lived in Shiraz, she went to perform pilgrimage at Shah Cheragh every week, and at Astuneh, another popular shrine.

When she visited Shah Cheragh or Astuneh, like the other women performing pilgrimage, Esmat held on to the grillwork surrounding the tomb and, murmuring her prayers, gradually circled the tomb. People would shove donations into slots through the plastic walls on the other side of the metal grillwork. While at the shrine, Esmat might perform her evening prayers and sit for a while by herself, or with her female relatives if she had come with them, murmuring prayers while fingering her *tasbih* (rosary-like string of prayer beads). In the shrine and outside in the courtyard, small groups sat together on the ground, relaxing and conversing. During holy days and on days of the week considered particularly efficacious for undertaking pilgrimage to a particular shrine—Sundays for Astuneh and Thursdays for Shah Cheragh—the shrine areas would become very crowded.

I remember when a delighted Esmat first traveled to Mashad in northeastern Iran, in the company of her husband's family, on a pilgrimage to the tomb of Imam Reza—brother of Shah Cheragh, the only Shi'a Imam buried in Iran. By the time I returned to the village over twenty years later, Esmat had been able to make the pilgrimage to Imam Reza in Mashad several times.

For a woman who was restricted by a watchful family, relatives, and neighbors, and constrained by lack of resources and modesty requirements even more stringent than those for other village women, since she was a widow, the Shi'a Muslim practices and stories provided an outlet for Esmat's spirituality and imagination, and for her thirst for knowledge, meaning, social interaction, mobility, and experiences. Particularly, owing to the lack of other opportunities, religious events and activities formed the frame on which Esmat constructed her life.

As mentioned earlier, Esmat married when she was fourteen. Her husband died four years later in an accident. He had been working at the Post Office and Communications Department delivering mail by motorcycle for only about a year when the accident happened. When her husband died, her son Karim was two years old and she was about forty days pregnant with Reza. She went to the Post Office and Communications Department for assistance. But they said that her husband was not an official employee. Therefore they wouldn't give her the pension for wife and children.

Esmat had to go back to the village and stay with her husband's mother and father. She lived with them until Reza was born and turned a year old. It was hard to live with them. They didn't treat her as a daughter-in-law, the wife of their son who had died. She felt as if she were working for them, washing their clothes and doing housework, so that they would give her and her sons food and clothing. She felt like a servant, so she left and went to stay with her mother. There, she got a sewing machine and taught herself to sew. As Esmat would say, first God, and then her mother helped her. Her mother bought and sold cloth and took care of the children so that Esmat could sew.

The Bicycle Accident

When her older son Karim was twelve and Reza was ten, they were involved in an accident. They were both riding on the same bicycle when a car hit them. Reza was hit on the leg and the arm, and Karim was hit on the head and had internal bleeding in his brain. He was unconscious for eighteen days. His hip was also broken. At the hospital, the only thing that could be done was to pray. Esmat stayed there praying. But then she made a vow in that hospital so that her son wouldn't die. She vowed that if Karim survived, she would send him to the front lines of the Iran-Iraq War.

Karim was in very bad shape one night. The doctor was doubtful whether he would survive. The doctor said that even if he did survive, he would be deaf and dumb or paralyzed. After the doctor said this, all that Esmat could do was to stay at her son's side praying. As she later told me, all she could do was to trust in God.

Late that night while she was sitting at his bedside, crying and praying, a man came into the room. He seemed to be a very good man and asked her, "Why are you crying?" Esmat answered that she had raised her son and now he was in this state. The man said, "Say this ten times: O 'Ali, I swear by the Name of the Imam-e Zaman. O Imam-e Zaman, I swear by the head of 'Ali. Say this just as I have said it to you." Esmat recited this over and over and over until morning. He had asked that she say it ten times, but she had recited it hundreds of times.

In the morning Karim regained consciousness. His hip was broken, and he was in a cast from his neck to his leg. Everyone had a different opinion about when his hip would get better. They opened up the cast in three months and his hip was healed. The doctor said this was a miracle, that only the Imams and Christ could perform miracles. Karim, who was in the room and listening, said, "Maybe it is because you said it so many more times than he told you to that I recovered so well."

For six months Esmat worked to help her son get better. Twice they said that his hip was crooked and that it was not healing right, so they broke it again and reset it. When they wanted to do that a third time, she refused. She walked with him while he held on to the bars, and she massaged his legs. Finally he went back to school.

When Karim was sixteen, he came to her and said he wanted to leave school and go to fight in the war. Esmat remembered her vow. She told him to go since she had promised that if he survived she would send him to the battlefront. He went to fight the war—the war between Iran and Iraq that lasted from 1980 to 1988.

A Son at War

Esmat spent her evenings and Saturdays at shrines praying that all the young men would return in good health from the battlefront, and she prayed the same for her son too. She had gone through so much trouble in raising him. As for Karim, he was in training for about two or three months, and then he was at the war front for eighteen months. He was posted in Shalamcheh, Mahabad, and Dezful on the southern battlefront, and he also fought on the northern battlefront. He served on the Faw Peninsula too.

He would come home for a week every three months and then go back again. If he had stayed on in the army, Esmat thought, he would have become a colonel. But Karim didn't want to continue in the army.

Meanwhile, Esmat kept to herself. In the summer, she didn't seek out cool places, saying to herself, "My son is in hot places." She never went outside to sleep at night. During the cold weather in the winter, she didn't come up close to the stove, saying to herself, "My son is in cold places." She never went any place for fun. She didn't even go to visit her relatives. She just worked and then went to the tombs of the imamzadehs to pray for her son.

Once she noticed that a crowd had gathered outside her house. They said that there had been an attack at the war front where Karim was serving at the time, at the Shalamcheh front on the Iraqi side. Many Iranians had been killed. Esmat's mother began crying. Esmat asked her, "Why are you crying?" She replied, "All of the young people have been killed." One person had come back from the battlefront and they asked him, "Did you see Karim?" He told them that Karim had gone ahead and then there was an explosion, after which he couldn't see Karim anymore. Esmat told her mother not to be unhappy. "After all, Karim doesn't belong to us," she said. "We shouldn't be unhappy. Whatever God wants."

Then she went to Shah Cheragh and prayed. She prayed to God for all the young people to be safe and for Karim to come back safely. She went to the hospital and visited a young man who was wounded. His arms were gone. She sat at his bedside, laying her head near his head, and cried. It was as if she were sitting and crying at the bedside of Karim.

She always told people that Karim wrote to her and phoned her. She knew that in secret they said that she was lying. She later stated that she always thought that he would come back. She never thought that there was a possibility she would lose him.

She also held *rozeh* in her home. On Sunday evenings, the people who went to the mosque to pray came to her house afterward to say the special prayers of request. They always went to someone's house after Sunday evening prayers at the mosque. On these evenings, some women read from the Qur'an and prayers were said. Women came in the afternoon and men at night. Many people came to these prayer evenings and all of them supported the Revolution. First they prayed, and after the initial prayers were over, they held a special prayer for Imam Husayn during which they beat their chests in mourning. Then refreshments were served—tea, cake, and sometimes cucumbers, or fruits such as apples. The people who attended these prayer gatherings formed a group. They had set up a library and an office and arranged for *rozehs* to be held. Other people came and listened. The prayer gatherings were held for the sick, for those who had sick ones or suffered unhappiness, and for those who were at the war front.

Esmat made a vow to Bibi Zaynabiyyeh, the sister of Imam Husayn. She prayed that she would offer a meal for Holy Roqayeh, who was only three years old when they brought the severed head of her father to her. "She was a little innocent child. God will listen to her, God will give her attention." Esmat vowed that she would also offer a meal for Abul Fazl. If Karim came back, she also vowed to distribute treats at the shrine of Astuneh. She would buy a lamb and offer the meat to the poor. She made all these vows knowing that if you got what you requested, you had to fulfil your vow, even if you have to borrow to do it.

Thank God, Karim wasn't wounded. On one occasion, he fell into water, and the water was about to pull him under. He said, "Oh, God, Ya Abul Fazl, if this happens, let my body be carried back to my mother. Don't let me be buried in Iraqi soil." He grabbed onto a rock and pulled himself out of the water.

People always came to Esmat and said that Karim was killed and that his body couldn't be found. When finally he came home, she told him not to go back to the war front again. But he said that he had to go back.

She prayed again. She prayed that things should be so hard for Karim, that he should be safe but there should be nothing for him, neither bread nor water. Karim returned to the war front. He wrote a letter to his mother and said, "I swear, when I come to get water, there isn't any. When I go to get my food, it is all finished." He wrote, "Mother, Pray for your accursed child."

Karim came back home again, and then the war ended. He hadn't yet reached home when the peace treaty was signed.

The Death of Her Mother

During the Iran-Iraq War, Esmat hadn't been able to sew because she was so anxious for her son Karim. She had gone to the Development Office and explained her situation. She had asked for work so that she could go on with her life. They sent her to work in a government childcare center for young children. The parents of these children were doctors or other professionals, and they took their children to the center to be cared for while they went to work. Esmat fed and cleaned the children. She met her friend Maasumeh there. Her life there was good, she said, until the childcare center closed. When she went to the Office of Employment for assistance, they gave her eighteen months of unemployment pay. With this money she was able to go to school. During those eighteen months, she attended adult education classes and studied Persian through fourth class in elementary school.

After that she opened a shop in ‘Aliabad, across from the gas station right along the highway, and sold cloth. She ran her shop there for three or four years. She would go to her shop in the morning and come back at night. She continued this routine until her mother got sick. When her mother became sick, Esmat couldn't leave her alone. So she had to give up the shop. Even after she gave up her shop, she continued to sell cloth at home while she tended to her mother. She continued to do this until she lost her mother. Karim had married while her mother was still alive, so they all had lived together, Esmat, her mother, and Karim's wife, for about eight years. Karim's son Akbar was born when her mother was alive. He was a year old when she died.

Her mother had a brain tumor, a very fast growing cancer. With the help of her brother, they took her to the hospital. They took her to the city of Isfahan for an MRI, which showed that an operation wouldn't help. She didn't get any better. In two weeks' time her brain stopped working, and she stopped recognizing even her own daughter. For one week she didn't recognize anyone, then she wasn't able to walk, sit, or hear. She couldn't even pray. Her only words were, “O ‘Ali, O ‘Ali.”

Esmat was terribly unhappy for two or three months after her mother died. She couldn't do any work. She said, “Oh, God, my only protector was my mother. You Yourself please help me so that I can go on with my life.”

Persistence and Trust in God, the Imams, and the Imamzadeh

Esmat reflected on her life, emphasizing how much God had helped her. She recalled that she didn't know anything about sewing when her husband died. But God had helped her so that she could learn to sew and thereby earn an income to support her children. This way she didn't have to ask for help from anyone. According to Esmat, if someone sits at home and doesn't work and expects someone else to help, it defaces God's reputation. A person must make an effort and work to support oneself. According to her, real Muslims will think along these lines. Indeed Esmat had made that effort. She acknowledged that foremost God had helped her, and then her mother. It was very difficult for Esmat when her mother died, and even today she doesn't find it easy. But according to Esmat, when we lose someone, we say, "God, give us patience."

When her husband died, she said, she didn't even know what religion was. All she wanted was help from God to do a good job bringing up her children. She asked God to help her raise her children without any help from her in-laws. Her mother told her to ask for help from Imam 'Ali, the son-in-law of Prophet Muhammad and the first Imam or leader of the Shi'a community after the death of Prophet Muhammad. He was known as the father of orphaned children. (In the Middle East a child who has lost the father is considered an orphan.) Imam 'Ali had provided food to orphans for their dinner. "Ask Imam 'Ali for help in raising your children, so that you don't have to ask anyone to help you," she had told Esmat.

After that, Esmat got started. On her own, she practiced sewing and learned it so well that everyone came to her to get their clothes made. As she put it, her religion helped an illiterate, eighteen-year-old widow to stand on her own feet and raise her children, and to be both father and mother to them. She had done a good job of bringing them up, too, she said. Many young men in the village were drug addicts, but her sons were not among them.

When I left Iran after my initial research, Esmat recalled I had given my household belongings to her. Other people had asked me for these items, but I had promised to leave them with her. She didn't want any help from local people, but I was a stranger. Yes, she thought that I was kind, but that this kindness had come from God. It was God who made me realize she needed those things.

When she couldn't sew anymore, she attended a course in how to use a knitting machine, and she knitted things to sell. Then she worked at a daycare center for eight years. Her mother was with her for as long as she needed her. Her son Karim had married. She had opened the shop. Then God had taken her mother from her, leaving her so unhappy for a year. When she lost her mother, she thought that she wouldn't be able to work anymore. She felt that she had no one in the world, and that, except for God, she was all alone.

But again she asked God to help her so that she could resume work. She didn't lose her spirit She made efforts so that her life could go on again and she could keep up her good name. She had trusted God through all the unhappiness in her life. Esmat noted that we all have God in our religion, including the Muslims, the Jews, and the Christians. Only the Communists don't believe in God, she said. God had helped her. She recited the prayers of intercession to the Imams and the fourteen Ma'sum (the Innocent Ones) and asked for their assistance. They had helped.

Esmat preferred to rely on God and the holy ones not only because she believed them to be more powerful in their ability to assist her, but also because assistance obtained from them would not put her in any vulnerable position. Openly revealing what she really thought and felt, or asking earthly people for help, could be dangerous. People might use her confidence and dependence on them in ways that could harm her. Other than her mother, the people from whom Esmat had sought assistance had turned out to be unreliable. When she needed assistance from her sister-in-law, so that she, a young bride, would not be alone at night, the sister-in-law used this opportunity to cause trouble between Esmat and her husband. As a young widow, when Esmat turned to her in-laws for support and a place to live, they treated her like a servant. When Karim was in a terrible accident and unconscious in the hospital, instead of offering her emotional support, her brother-in-law blamed her. "See," he said, "You went to marches (in support of the Islamic Republic), and look what happened to your son." When Karim was at the battlefront in the Iran-Iraq war and she was living under a cloud of fear, other villagers asked her whether Karim had written and whether he had called. They only wanted to see how she would react, perhaps to see if they could elicit any signs of fear or suffering, she felt. When she said he had written, thereby putting on a pretense of confidence and assurance of his safety in front of others, they had talked about her behind her back. She hid her fear and raw grief from others in Aliabad. Instead, she confined her self-revelation and desperate pleas to God, the Imams, and the imamzadeh. It was far better, she felt, to share conversation and community with God and the holy ones, whom she could trust.

Finding the human community of 'Aliabad to fall short, Esmat created a spiritual and social community for herself among God, the Shi'a Imams, the imamzadehs, and the Fourteen Innocent Ones.

PART II

In a New Place

CHAPTER 5

Emine: Muslim University Student in Berlin *(Turkish Student in Germany)*

Katherine Pratt Ewing

I first met Emine in a Berlin public library in Germany in the fall of 1999. Neatly dressed in a long skirt, light raincoat, and color-coordinated headscarf with delicately crocheted edges, she was looking through a rack of videos. I had come to this library, located in the heart of Berlin's large Turkish community, in the vain hope of finding a specific Turkish film, *Yalnız Değilsiniz* (You Are Not Alone), which is about the tribulations of a young Muslim woman in Turkey who had decided to adopt the headscarf in defiance of her secularist parents and was barred from classes by her college professors. I approached her and explained in my halting Turkish and German that I was a professor from the United States, and asked her if she knew the film. Trying to explain that I needed to know the name of the director of the film for an article I was writing, I finally gave up and resorted to English, which she said she spoke a little. She not only had the film, but also offered to lend it to me. We arranged to meet again at the library a couple of days later, and she handed me, a complete stranger, a bag with two films and a couple of Turkish novels that she thought I might be interested in. Much later she told me that she had been at the video rack on that first day looking for American videos to improve her English, and that our encounter was the first occasion when she had ever spoken English outside the classroom.

Over the next few months, which have now stretched into years as I return each summer to Berlin, she gave me many hours of her time, tirelessly helping me with my research on Muslim religious practices in Germany in the midst of her own busy schedule as a university student. She came with me as I visited many of the Turkish mosques in the city, always asserting that it was her duty to help me in my work. She and her family opened their lives to me, so that I could understand why they have chosen to live as Muslims in a non-Muslim country and, in turn, help to communicate to others what it means to be a Muslim woman today.

Emine's Family and Growing Up in Berlin

Emine grew up in Berlin. She has a family background that positions her as one of many thousands of second generation immigrants whose parents came from rural Turkey in the 1960s, 1970s, and 1980s, when West Germany was in need of labor to sustain its post-World War II economic miracle after the Berlin Wall had stopped the flow of workers from East Germany. Emine's parents came to Germany from Giresun, a small town just west of Trabzon, on the Black Sea in northeastern Turkey. Emine's mother Fatma had accompanied her guest worker parents to Germany as a teenager, and when it was time for her to be married, her marriage was arranged with a cousin back in Turkey. Her father was studying to be an Islamic scholar but came to Germany to join his wife immediately after finishing his studies and before he could begin teaching in an Imam Hâtip, a high school in which Islamic subjects are also taught. When he did finally join her in Berlin, he took a job in an automotive factory, where he still works in alternate day and night shifts on an assembly line. After twenty years in Germany, his ability to communicate in German remains limited in comparison with his eloquence in Turkish (though he says that after being away from Turkey for so long, he finds it difficult to read recent scholarly works in Turkish because the language has changed so much since he left). Emine's mother worked as a cleaning woman at the Free University in the early years of their marriage but gave up her job when her fourth child Elif was born in 1989. She claims to have forgotten the little German she learned while she was working, though one day when I came to her house after speaking with Germans all morning, she tolerantly switched to German when I found it difficult to make the transition to Turkish.

Like most Turkish families in their situation, Emine's parents spent the early years of their marriage expecting to return to Turkey one day. They made a minimal investment in their life in Germany, sending some of their earnings back to family in Turkey and saving the rest. In Berlin, they lived in a two room flat that they furnished with used furniture, the cheapest they could find. The flat was in Kreuzberg, an area of Berlin that is often called "Little Istanbul" because of its high concentration of Turkish immigrants. The three children in the family slept in the living room, and they often got sick, suffering from allergies caused by dust in the furniture.

The family lived like this through Emine's elementary school years. Many children in the school were Turkish or other minorities. Most of the Turkish speaking children were segregated into separate classes because of their lack of fluency in German. Emine's parents insisted on a regular class for their children. Nevertheless, her parents tried to ensure that their children learned Turkish, by selecting it as their second language in school, so that they would have no problems and be able to return to Turkey.

Because both parents had jobs and worked in shifts that were not usually compatible with the needs of their children, Emine and her two brothers often had to care for themselves. They had to shoulder the responsibility of getting themselves off to school in the morning: "When my twin brother and I were six, my mother was working many hours, and my father had to go back to Turkey to serve in the military for a year and a half. We got up alone with an alarm clock. My mother made our food the night before. On the day before my father left for Turkey, he showed us how to travel on the subway by ourselves. He taught us to read the 'O' in Osloer Strasse, the name of the direction we had to travel. Normally it is forbidden for children to travel alone at that age."

And things sometimes went wrong. It is the custom among Turkish families, and among some German families as well, that people remove their shoes at the front door before entering a home. Emine's family keeps a shoe rack just inside the door, and there is a row of slippers in the hallway for people to don once they take off their shoes. One morning, when she was about ten, she and her brothers headed out the door to walk to their school. It was a wintry, snowy day, and only when she reached the street did she realize that she had forgotten to exchange her slippers for shoes when she came out of the house. The door had locked automatically when the children closed it after them, and they did not have a key to get back in. Emine had no way to get her shoes or to reach her mother, so in desperation she made her way to her uncle's home a few blocks away, feet nearly frozen, and spent the day waiting there until one of her parents got home.

As Emine told me these stories, her mother began to cry gently regretting that Emine and her brother had been latchkey children. When she was very small, Emine had to care for her two brothers; a duty that she feels has shaped her personality and made her very serious and often worried. Emine remembers worrying a lot, and her mother feels that now Emine experiences too much stress because of all the responsibility she had as a child. Emine draws sharp contrasts between her own childhood and that of her younger sister Elif, who is about eleven years younger than her, and she links these differences to the family's changing situation as they became more established in Berlin. While Elif was growing up, her mother no longer worked. Elif had a far more relaxed childhood, and, as a consequence, Emine feels she has a more cheerful personality.

When Emine was about eleven, it was time for her to begin wearing a headscarf. As she described it, "At the beginning of the fifth class, I started to wear the headscarf. I was afraid that my teacher would tell me to take it off, because he had done that to another student the year before. After the first lesson began, a German boy asked the teacher if I would always wear the headscarf. The teacher said, 'Ask her yourself,' but he never asked me—he didn't dare. Actually this subject was discussed very little." The difference marked by the headscarf was surrounded in silence.

I have been able to observe Emine's younger sister Elif go through this transition of wearing a headscarf. When I first met the family, Elif was a very active ten-year-old interested in karate and running, and she usually wore sweat pants and other casual clothes to school. In the summer when she was between the fifth and sixth grades, she decided that she would begin to wear the headscarf in sixth grade, but after a few weeks she changed her mind and decided she wasn't quite ready for it, in part because she still had too many short-sleeved shirts that she hadn't yet outgrown. When I saw her again near the end of the seventh grade, she was as lively and dynamic as ever and had begun to wear the headscarf and long skirts, though she was not yet expert at maintaining proper decorum. Elif explained that she had really wanted to begin dressing like this and now felt completely ready, but there was a casualness about her that I doubt Emine ever displayed. As we sat around the coffee table one afternoon, right after Emine had gotten home from school, her mother adjusted Elif's scarf to cover a small area that had become exposed above her scoop-necked jersey.

Throughout Emine's elementary school years, the family had always planned to return to Turkey eventually. But, as happened with many of their friends and relatives, there came a time when they realized that this wish would never be fulfilled. It was a significant turning point in their lives. In 1991, when Emine and her twin brother, who were the oldest children, were about twelve, her father made arrangements for the family to go on hajj, the pilgrimage to Mecca. Before the trip, he announced that after hajj they would move back to Turkey. He felt that if they didn't do it then, they would never return. He was especially concerned that once the children were teenagers, it would be very difficult for them to adapt to the Turkish educational system. But at the last minute they decided that it was already too late, and the family ended up staying in Germany.

Once they knew that they would remain in Germany for good, they decided to invest in their future in Berlin. They rented a bigger apartment, which had three bedrooms, a kitchen, and a large living and dining room. They bought new furniture, including leather sofas, a marble coffee table, and a wall of bookcases and cabinets. Several years later they took advantage of a government program that made it possible for them to buy their own apartment. Finally, they made sure that their youngest daughter Elif, who was considerably younger than the other three children, studied English rather than Turkish as her second language when she reached the fifth grade. Though Elif studies Turkish in a course offered by the Turkish consulate at the school after school hours, this decision meant that she would be unlikely to learn enough written Turkish to be able to function as an educated professional in Turkey. But she would likely become proficient in English more readily than her sister, a skill that will enable her to travel all over the world. Emine is dedicated to becoming fluent in English, but she feels that she has been at a disadvantage by beginning so late.

The presence of so many Turkish guest workers and their families in German cities has led to difficulties with their integration into German society and problems with discrimination and negative stereotyping by Germans. Though Turkey's big cities are very cosmopolitan, most of the guest workers came as unskilled laborers from rural areas and had little education. The "culture shock" of this transition was powerful. When they brought their families to Germany, most of them were concerned with maintaining the practices that had been the foundation of family and village life in Turkey. These included the separation of men and women in most activities and the importance of the multigenerational extended family. In the extended family, financial resources were usually shared, and elders had extensive authority over their children and grandchildren, including the authority to arrange marriages for their children. Gender segregation was often exaggerated by them in the German cities, because the men from Turkey were concerned with protecting their wives and daughters from what they perceived as the threats of Western society. When experiencing the stress of a difficult job situation and social discrimination, some fathers became more authoritarian than they might have been back in Turkey and justified their authority in terms of Islam. Many children growing up in such households chafed at the constraints placed on them by their parents and felt that Islam as practiced by their parents was irrelevant for their own lives.

Emine's family did not follow this pattern. Though they had migrated as guest workers from rural Turkey, Emine's father was highly educated, and had intended to become a teacher in an Islamic school. Islam continues to be a central part of the lives of everyone in the family. The family relies on Emine's father as a source of religious authority as they face the complex questions that often arise when they seek to maintain proper Islamic practice in a non-Muslim society. But they do not consider that they are simply trying to preserve their traditions while living in Germany. On the contrary, they question what they call traditional Turkish culture and the customary practices that many Turks in Germany have brought from their villages. Emine feels that because most guest workers and their families had received no Islamic education before they immigrated, they often confuse these customs with true Islam. She is critical, for example, of what she considers superstitious village practices such as wearing an amulet to prevent illness. She emphasizes that men and women are equal in Islam, and that a father does not have the right to prevent his daughter from getting an education or to force her to marry a man against her will, as, from her perspective, sometimes happens in villages.

Education has been a major focus in her family for both boys and girls. Although Emine just missed getting grades high enough to attend gymnasium, the highest level of high school, which prepares students for university programs, no doubt because of her language difficulties and the segregation of students in the early grades, she did well enough in the comprehensive school and on entrance exams to attend university. She had hoped to become a medical doctor, but her exam marks were just short of the cutoff for that year to be able to enter a medical program right away. She enrolled in Berlin's Free University and decided to become an elementary school teacher. Emine's success at getting into the university was a powerful marker indicating how much the family's position had changed since their early years in Germany. Fatma's voice was filled with emotion when she said to me, "I never imagined when I was cleaning toilets at the Free University that my daughter would one day be a student there." The German government pays the educational and living costs of university students, which means that for most students who want to continue studying and are academically qualified, there are few financial obstacles, especially for those who live at home while studying, as Emine had done.

Despite their decision to remain in Germany, the family maintains close ties with Turkey. Like most Turkish families in Germany, they continue to spend several weeks in Turkey every summer, sometimes flying there and sometimes making the long trip from Berlin by car on often treacherous stretches of road across several countries. They would return to Germany with big bags of hazelnuts picked in their family's orchards. Among the friends and relatives they visit in Turkey are Emine's grandparents, who moved back to Giresun from Germany after her grandfather retired.

Reflecting on her family's migration, Emine once told me: "When I think about the situation in Turkey, I tell my parents sometimes, 'It was good that you came to Germany.' I could sometimes imagine myself going to Turkey in order to try out life there, because Turkey is actually a very beautiful country. But I would only do that if I was certain that I could return to Germany. Everything does not function as well in Turkey as it does in Germany, for example, the health system and the school system. In Turkey it is more difficult to study at the university. And studying with a headscarf would be a problem."

Describing what life is like in Berlin for a Muslim whose family is from Turkey, Emine said: "I like being in Berlin. I do not hear slurs on the streets very often, though it does happen once in a while. Sometimes funny things do happen. When I was in the sixth grade and went to the career information center in order to get information about a career in the pharmacy, I received a harsh answer. 'No, you mean a pharmacist helper.' Or when I went to the Deutsches Theater with friends, an old woman sitting behind us said to her friend, 'Look, they can speak German.' I asked myself, what would we have been doing at the theater if we couldn't speak German? But I'm happy in Berlin. Maybe that will change if I do not receive a job."

Unforeseen Tragedy

After I came to know them, the family was struck by a shattering tragedy. I first heard of it when Emine called me in the United States shortly before she was to visit me, a trip that we had planned with great excitement. Her voice shook as she told me she wouldn't be able to make the trip because her twin brother Muhammed had just been struck and killed by a truck while riding his bicycle to the university. They were going to fly the coffin back to Turkey immediately and spend three weeks with relatives. The family had a difficult time coping with this unimaginable catastrophe of losing their eldest son. Before the accident, he had recently returned from a five-week trip to Malaysia and Indonesia, where he had stayed with Muslim host families and marveled at how differently Muslims could live. I learned of these details the following summer when I was back in Berlin and with Emine and her mother; we were then going through photo albums, looking at the many pictures that Muhammed had taken on his trip.

They shared with me the devastation they had felt when they heard of the accident and broke down as they reminisced about his energy and his plans that had been so abruptly and totally ended. But I also saw how they drew on their relationship with God to help manage their loss and how their understanding of Islam shaped their experience. Before Muhammed's death, the family would read from the Qur'an after evening prayers once a week on Thursday, which is the evening of the Sabbath. After Muhammed died, they recited one of the shorter chapters from the Qur'an every evening. Elif, who was learning to read the Qur'an from the imam at the mosque that her family attends, recited her portion from memory.

Emine's efforts to absorb this loss and to render it meaningful can best be communicated in her own words, which she wrote to me as an e-mail message:

> It is now more than two months ago that my brother died, but we miss him every day more and more. It is very hard. OK, everyone will die, but if the person was young and had her/his life before them, it is much harder. My brother had so many plans, he liked traveling and meeting different people. He planned to found a firm after his school. He wanted to help people. He wanted to learn more about Islam and also tell people about Islam. He had also done this, but he wanted to do more. At that time he had not had time to do much because of school. How could he know that he would die so early? If we think such things we go crazy. But on the other side, we think that he left the world before gathering many sins. Could you guess what he said two or three days before he died? He said, "To die young is good on the one side, you die before having many sins." I or my other brother replied, "You can also think the other way; if you die early you can also

not do many good things (*savab*)." Could you imagine, we had this conversation two or three day[s] before he died, among my two brothers, my mother, and me. I had forgotten this conversation, but my mother reminded us.

Some events and conversations we understand better now. Also some dreams. For example, one month before my brother died, my mother dreamed that she gave something (she thinks it was bread) to her dead grandparents. My mother had heard before that giving something to dead persons is not good, but she didn't have misgivings, because her grandparents were smiling in the dream. And my father saw in a dream some days before the accident his sister who died when she was fourteen years old. In his dream she was giggling in her grave. Now we can understand these dreams better. The dead persons were happy that my brother will come to them.

We buried my brother next to my father's sister, because she is our nearest dead relative. I had never seen her; she died one and [a] half years before we were born. Not anyone from my nearest relatives or friends has died that I remember—only one aunt of my parents died last year. (My parents are cousins.) All my grandpas and grandmas are living, even the parents of one of my grandpas are living. Then suddenly my brother died. I had also never seen a dead person before. Muhammed's best friend died four years ago of a tumor (cancer). Muhammed was very sad about that, but if my mother cried (we all knew his friend) Muhammed said: "Why you are crying? He is my friend, and on the other side, do you think that dead people are really dead? No, they are living but we cannot see them." And my brother was not afraid of the dead. But he could also not know that he would die young. Now we see him only in our dreams.

Later, Emine told me of a dream that she had the night before Muhammed died. "On the night before he was killed, I dreamt about a news report: In the news I could see a long red subway and a voice reported that ten people died in a subway crash. The voice in the news report was so urgent and pressing that I could hear it even after I woke up." According to Islamic belief, dreams have always been understood to be an important window into the unseen world.

As the children grew into adulthood, it became more difficult to get the whole family back to Turkey, especially after Muhammed's death. For the first time, in the summer of 2003, Emine's parents took only their youngest, Elif, with them on their trip. Emine and her younger brother Yusuf stayed alone in Berlin because of the demands of their university programs. Emine was working on a major paper, one of the final requirements for her degree. But she also said that the idea of going back to Turkey made her very sad because it reminded her of her twin brother, of the times when her family had been together on vacation at their summer cottage that they had recently bought in Izmir. It was a site they had chosen because the developer had promised that there would be a separate beach for women, which would allow them to wear bathing suits and swim without worrying about their modesty being compromised.

To Be a Good Muslim

One of Emine's basic orientations is to strive to become a model of the true Muslim and to teach others about Islam. What does being a good Muslim mean for Emine? In addition to following the five pillars of Islam, she is always aware of being an example to others of what a good Muslim should be, following the example of the Prophet Muhammed. For Emine, wearing a headscarf is part of this responsibility. She told me how one day on the metro she saw two girls with headscarves who were very conspicuous because they were talking very loudly and laughing. Emine was embarrassed because of them. She asked them to be more silent, but they just ignored her. Other girls violate the principles of Islam when they wear pants and makeup along with their headscarf.

Praying regularly while attending school in Germany has been challenging: "During school I observed prayer times. In the school there was a hall to the computer room which was usually empty. A friend watched out for me and I prayed. Then I watched out and she prayed. Prayer was a problem in the school. Even now at the university, it is not easy. There are no prayer rooms. In order to wash, we use the handicapped toilets and to pray we use the stairwell."

In addition to her university courses, Emine has been taking a correspondence course on Islamic education. The issue of Islamic education has been a controversial one in Berlin. In contrast to the United States, where there is a clear separation of church and state, so that public schools must strictly avoid sponsoring religious activities, many German states include religious instruction as part of the public school curriculum. In Berlin, students until recently were able to choose only Catholic, Protestant, or Jewish instruction. Many Muslims felt that their children should also be entitled to receive religious instruction in schools. A few years ago the Islamic federation, a Berlin organization with which Emine's mosque is affiliated, sought to develop a program of Islamic education. When the Berlin public school system blocked them from implementing it in the schools, they filed a law suit, which they finally won in Germany's Constitutional

Court after several years of litigation. As this opportunity for Islamic education has opened up, so, too, has the need for teachers. Emine, along with a few other young Muslim women who had already decided to become teachers, began studying how to develop an Islamic curriculum through a correspondence course based in the German city of Bonn. She periodically traveled to Bonn with this group of young women for weekend programs associated with the course. In addition to studying Arabic, they learn an array of subjects. For example, one of the seven volumes of the course is devoted to Islam and the environment; it stresses that an important part of being a good Muslim is protecting the environment, which includes learning how to recycle correctly, following the German recycling codes that are printed on all packing material. The curriculum places great emphasis on practicing Islam in a modern society.

When I returned to Berlin in the summer of 2001, the summer after Emine's brother's death, I raised again the possibility of her coming to visit me in the States. Emine told me that it wouldn't be possible because her mother would be too upset, and she would worry about having Emine so far away from home after losing her son. But as I said farewell at the end of the summer and reminded her that she was always welcome to come for a visit if she could, she told me that she had been talking it over with her parents, and they had agreed that it was a good idea after all.

Emine was scheduled to arrive in North Carolina just ten days after what proved to be a world-transforming event, the plane hijackings of September 11, 2001. She e-mailed me on September 12 to express her sadness for the victims as well as her anger at people who "cannot be real Muslims," if they commit such acts of violence. She wrote that her mother was particularly upset because Emine "could have been in one of those planes or near one of the buildings," clearly a reference to her planned trip to visit me. I was sure that once again events had conspired to prevent her trip, but after a few more e-mail exchanges, in which I assured her that people at Duke University were very supportive of the Muslim community and that we would not travel to a big city such as Washington because of the possible danger, she courageously decided, with her parents' support, to come after all.

While in North Carolina, Emine was particularly interested in learning more about the local Muslim community and about how Islam is taught in an American university. She sat in on classes at Duke University, visited a Durham mosque, and attended meetings of the Muslim Students Association on Duke's campus. She took photos of statements of support for Muslims written in chalk on Duke's sidewalks in the wake of post-September 11 reports of violence against Muslims in the United States and was amazed at the sight of the American flag flying everywhere.

Career Challenges in Germany

In the fall of 2004, Emine completed her undergraduate work at the Free University, where she studied to be an elementary school teacher. As she wrote her final thesis, on the topic of the role of education in social inequality, she worried about her job prospects. She already had difficulty finding a job: "In the last year of school I was assigned an internship at an elementary school. Before classes began, I reported to the head of the school in the office. He said to me, very casually, that I must remove my headscarf. 'Unfortunately I can't do that. This is required by my religion,' I answered. Then he spoke with my university professor. The public school office got involved. The result was that I was not allowed to do my internship at that school. The school principal kept saying that I would also not be allowed to teach in Turkey with my headscarf. Many Germans love to talk about the lack of democracy in Turkey, but when it comes to the headscarf, they often use the restrictions on the headscarf in Turkey as a model of how Germany should be." Emine was able to complete the internship requirement only because she was lucky enough to obtain a position at the only Islamic school in Berlin. This is the school where Fereshta Ludin currently teaches. Ludin has become famous throughout Germany because of her own struggle to teach while wearing a headscarf. After being fired from a teaching position in the state of Baden-Württemberg, Ludin filed a series of lawsuits and appeals that culminated in a suit before Germany's Federal Constitutional Court, which was decided on September 24, 2003. According to that decision, the government of Baden-Württemberg had been wrong to fire Ludin because there was no existing law against wearing a headscarf in that state. Though Ludin appeared to have won the case, this decision upset the Muslim community very much, because instead of placing the Muslim woman's right to wear a headscarf under the protection of the constitutional right of religious freedom, it opened the door for state governments to pass laws explicitly banning public school teachers from wearing headscarves. In the meantime, public controversy surrounding the issue made it more difficult for students like Emine to teach in schools where just a year or two earlier school principals had been allowing them to complete their internships.

When Emine completed her studies in the fall of 2004, she applied for a teaching position; by then, laws against teaching wearing a headscarf were in place in several German states. She and the other young women in her position then became politically active. She says: "Maybe I will have to move to another state in Germany, or I may have to try other work, for example, in the area of social work, which I would very much regret. My mother is already very sad that I have studied for so long and will maybe not receive a position. Trust in oneself is very important. The fact that I know that what I am doing is not wrong helps me to overcome the difficulties. Without strong conviction, it would be much more difficult for me. Belief in God and trust in God are very important for me."

CHAPTER 6

The Intricacies of Being Senegal's Lebanese Shi'ite Sheikh *(Lebanese Religious Leader in West Africa)*

Mara A. Leichtman

Sheikh Abdul Monem El-Zein is an imposing figure in his long gray robes, white turban, and gray beard. He is a charismatic man, and when he speaks, people listen. He demands respect not only because he is the first Shi'ite sheikh (Islamic leader) in Senegal, but also because he is one of the few highly educated men among the Lebanese community in Dakar. Here, knowledge is something to be shared, and Sheikh El-Zein is certainly not a man short of words. He lectures at more than four hundred occasions a year—weekly at Friday prayer, every evening during the month of Ramadan, and twice a day during the ten days of the Shi'ite holiday of *'Ashura*. On Tuesdays and Thursdays he teaches religious classes for men, on Saturdays he teaches Qur'an classes for women, and he speaks at memorial ceremonies and various other community events. He lectures about religion, politics, history, moral issues, and the daily struggles the community might face. For example, his theme for Ramadan 2002 compared the Qur'an, the Torah, and the Gospels. He has encouraged the community to live life to the fullest, to be unafraid of death, to seek knowledge and education, and to quit smoking. He has used such occasions to talk about the dangers of the Internet, which contains good sites about religion, but also has bad sites about sex, which children should be forbidden from accessing. He has chided the community for gossiping too much about each other, and once he gave a lecture against abortion. A woman who was pregnant with an unplanned child, and thinking of aborting, attended the sheikh's lecture by chance and decided to carry the baby full term. Such is the power of a sheikh.

Abdul Monem is also an open man. Deciding to work with an American Jewish anthropologist after 9/11 and during the tense time of the American-led war on Iraq is not a decision every sheikh would make. On my first day in Senegal in the summer of 2000, a community leader whom I had met the month before in Lebanon picked me up from the airport, helped me settle in with a Lebanese family, and sat me down in his office to ask me what the real purpose of my visit was. He told me that people in both Lebanon and Senegal questioned whether I was a spy. I did my best to assure him that I was not. I furnished him with a copy of the letter of support from my university and told him that my family was from Michigan. Detroit and its suburbs, especially Dearborn, is home to the largest Lebanese community outside Lebanon (approximately 275,000 Arabs with a Lebanese majority in 2000). While most people I encountered abroad have never heard of Michigan, all the Lebanese, both in Lebanon and in Senegal, know where Dearborn is, and many have relatives there. The fact that I am from Michigan, and that my mother worked with a Lebanese woman

Figure 3 Sheikh El-Zein

born in a small village in Senegal, helped assuage this man's fears. In the end he concluded that I was too young (and too beautiful) to be a spy, and even if I were, it did not matter, since he had nothing to hide.

But the spy issue never went away, and only intensified after 9/11, during the war on Iraq, and as a result of the American government's plan to build a military base in Senegal. During the war, the American embassy in Dakar advised me to avoid public places, especially mosques, and suggested the possibility of changing my research topic; at the same time Sheikh El-Zein assured me that the safest place for me, should I feel threatened, would be his mosque.

Given the difficulty of my situation, as an American and a Jew, I was nervous the first time I met Abdul Monem. While I had met a few Shi'ite sheikhs in Lebanon, I had always done so along with Lebanese friends acting as escorts and translators. I knew I had to wear a veil, and that it was forbidden for a woman to shake the hand of a religious man, but this time I was going alone to meet a sheikh. When I started to introduce myself nervously in French, the language I use with most Lebanese in Senegal, Abdul Monem stopped me, addressing me in Arabic, and informed me that the director of the Islamic institute usually translates between French and Arabic for those who do not speak Arabic, but in my case his presence was not needed. And so I had to get by in very rusty Arabic, which I was not at all prepared for. He was patient and encouraged my less than eloquent language skills.

I was to meet Sheikh El-Zein a few more times in both Dakar and in Lebanon before my extended period of fieldwork in Senegal, When he asked me about my religion, he was suspicious, yet curious at the same time, and fired away questions about which of my parents were Jewish (both) and whether I had ever been to Israel (I had lived there from 1994 to 1995). He questioned whether the Israelis tried to control me, or solicit my support, and what ties they had with the American Jewry. I answered his questions honestly, and tried to break whatever stereotype images there were. I told him that I did not agree with Israeli politics, and that I hoped for peace in the Middle East. He asked a number of questions on comparative theology, which I answered with the knowledge I had, and he seemed fascinated that there were so many similarities between our two religions. He told me that I was the first Jew he had ever spoken with and that it was a strange situation for him. He asked me why I was not afraid to be with him, or to be in Lebanon or Syria (where I had just spent the summer), and I asked why should I have something to fear just because I am Jewish? It would be another year after that meeting in Lebanon before I would return to Senegal, a year after 9/11, and in that time Abdul Monem decided that he would work with me. I was free to attend all the activities of his mosque that a woman was allowed to participate in.

The Early Years of a Shi'ite Sheikh: Family, Education, and Migration

Abdul Monem El-Zein was born in 1945 to a Shi'ite sheikh father and a Sunni Muslim mother in a small town in the south of Lebanon between Tyre and Bint Jubail. His father had taught him the Qur'an in his early years before he began formal schooling at age six. Abdul Monem studied in two primary schools: one secular and one Maronite Catholic. He then moved to Beirut where he attended a branch of the prestigious Sunni Muslim al-Azhar University of Cairo. He graduated from this *kulliyat* (faculty), which had taught him according to the Hanafi school of law, one of the four legal schools of Sunni Islam. With this educational background from secular, Catholic, and Sunni Muslim institutions, he began his formal Shi'ite education.

Traditionally, a Shi'ite sheikh is trained at one or more of the Shi'ite holy cities, such as Najaf in Iraq or Qom in Iran. This is more of an exception today, as Saddam Hussein had severely oppressed the Shi'ites in Iraq. Many Lebanese Shi'ites, therefore, now remain in Lebanon for their clerical studies. Abdul Monem's father had done his religious studies in Lebanon, but he wished for his son the opportunity to be educated in Najaf in Iraq. At first Abdul Monem did not want to become a sheikh. He had liked math and science in school and wanted to complete his education in those fields, but he also did not want to displease his father.

In 1961, Abdul Monem moved to Iraq where he attended the University of Najaf from 1961 to 1969. There he studied Qur'an and the *hadith* (the traditions of the Prophet), as well as *fiqh* (Islamic law), theology, logic, philosophy, Qur'anic commentaries, and medicine. He began to like religious studies in Najaf. During this time, he returned to Lebanon to marry at the age of eighteen.

Shi'ite Islam follows a hierarchy of clerical leadership based on the superiority of learning, a system that began in the late eighteenth century. The *mujtahids* (those who could follow their own independent judgment based on *ijtihad*, religious decisions based on reason) became the religious elite among the Shi'a. Anyone who was not himself a *mujtahid* was required to follow the rulings of one who was. The practice of emulating a *mujtahid* is called *taqlid*, and the *mujtahid* became the *marja' al-taqlid* (reference point for emulation). The *marja'* under whom Abdul Monem studied was Ayatollah Sayyid Abu'l-Qasim al-Khu'i. He also studied under other *ulama* (Islamic clergy). When al-Khu'i passed away in 1992, Ayatollah Sayyid 'Ali al-Sistani became the *marja'*, and Sheikh El-Zein currently follows his teachings, although he has never met him.

Abdul Monem's (now Sheikh El-Zein) first trip to Africa was in 1969; he was sent there at the age of twenty-five to lead the Lebanese Shi'ite community of Dakar. He had not planned to go to Senegal. He had received his baccalaureate degree from Beirut in English, and at first he had wanted to go to Nigeria, which was an Anglophone country. But two years earlier, Musa al-Sadr, the legendary religious leader who fought for equality for Lebanon's oppressed Shi'ites, had traveled throughout Africa. Representatives of the Lebanese community in Dakar, the oldest Lebanese community in Africa, had asked him to establish the first Shi'ite religious center in Africa. These representatives later traveled to Lebanon to remind Musa al-Sadr of their need. So in turn Musa al-Sadr journeyed to Najaf in Iraq to find the right man for the job. There he met Abdul Monem and appointed him to go to Dakar. When Abdul Monem protested that Senegal was a Francophone country and that he knew no French, Musa al-Sadr promised him that he would be conducting his affairs in Arabic and did not need to be fluent in French.

It was not easy for Abdul Monem to adapt to the life in Africa. He did not know the languages or the cultures. The Lebanese there were not accustomed to following a religious leader, nor were they accustomed to paying the religious taxes that he needed in order to do his work. At first Abdul Monem planned to stay only for a year or two in Senegal, but the years kept passing quickly as his work developed.

The Situation of the Lebanese Community in Senegal

Before Abdul Monem came to Senegal, the Lebanese embassy had chosen six religious men, who had made the pilgrimage to Mecca, to conduct weddings and divorces for Lebanese Muslims. However, these were businessmen and not formally educated in religion. Sheikh El-Zein writes in the introduction to his first book in 1973:

> The idea to write this book was born after my arrival in Senegal. The religious situation of the Lebanese community was disastrous with regard to a great void in religious culture and jurisprudence. This

> culminated in a lack of religious learning and spiritual practices, where prayer, the most important, was not at all satisfactory. Young people rarely approached religious circles and did not worry about their legal obligations.... The majority of parents were not interested in these problems. Others...were at too great a distance to learn the laws of religion and the rules of practice so that the symbols could be put to action and not only words.

Through courtesy calls on homes, invariably over small cups of Lebanese coffee, discussions took place between the young Sheikh and the community on different problems in society. On the part of the community these problems were largely the worries of emigrants, including economic concerns, lack of ties to the motherland, and problems with integration in a new homeland where they became second-class citizens. On his part, Abdul Monem slowly succeeded in making the community more aware of religious affairs and their obligations. But he had to learn about the community itself first.

The Lebanese community of Senegal is heavily concentrated in Dakar's city center. The Lebanese reside either in apartment complexes or in apartments built above their shops. They have always lived separately from the Senegalese, and, during colonial times, from the French. They were known to live in buildings that had an intermediate level of quality and comfort—in comparison to the Senegalese huts or wooden shacks and French villas or colonial buildings. However, many Senegalese now live in concrete apartment buildings, and wealthy Lebanese and Senegalese do have villas by the sea.

Such an area, being primarily commercial, is fairly polluted. The entire city is covered in dust that is blown south from the Sahara desert in Mauritania. The dust then mixes with trash, empty boxes, discarded containers, food remains, and other refuse, which attracts many flies and mosquitoes. In fact, certain Lebanese welcomed me to their "dirty city," and the Senegalese government even asked the Lebanese who owned shops on Galandou Diouf street, one of the main Lebanese-dominated commercial streets, to help clean up the city. Galandou Diouf is always alive with activity. The street is crowded with cars, motorcycles, carts with merchandise, and people who are shopping, selling items, yelling, spitting, coughing, and greeting each other. In the shops there is a Lebanese proprietor, occasionally a few Lebanese employees or co-workers, and a majority of Senegalese employees, or sometimes Guineans or Mauritanians. A clamor in a variety of languages can be heard from the shops. The Lebanese speak Wolof, Dakar's primary language, to their employees and Senegalese clients, a mixture of French and Arabic among themselves, and French or Arabic with non-Senegalese clients, such as French, other Africans, Moroccans, and Mauritanians.

Shopkeepers range from small grocers to sellers of African cloth, imported European fabric, household items, plastic mats, and cosmetics. Stores may carry the proprietor's last name, or may have more creative names, such as "Cedars," after the Lebanese national tree, "Bed, Bath, and Beyond," and "Al Pacino's Dream Shop. While Galandou Diouf is a commercial street, the nearby street of Ponty is filled with Lebanese fast food, falafel, and shwarma (a sandwich composed of roasted lamb shaved from a turning skewer) joints, and finer Lebanese-owned French bakeries and cafés. Senegalese peddlers crowd the street with their clothing, crafts, newspapers, and music. In another nearby area of Dakar, framed by Avenue du President Lamine Guéye and Boulevard de la République, the Lebanese own more expensive boutiques with imported European women's clothing and shoes, children's games and toys, jewelry, and the largest bookstore in Senegal. The Lebanese also own a few modern medical clinics. Whereas the first generation of Lebanese were primarily traders with little education, the second and third generations have begun to move into industry and the professions. They dominate the plastic, paper, and cosmetics industries, and even hold a share in the African cloth industry, manufacturing some of their own textiles. Among the Lebanese are also doctors, lawyers, dentists, pharmacists, tailors, and mechanics.

Bringing Lebanese Muslims Back to Islam: The Sheikh's First Major Accomplishment

Abdul Monem has brought formal Islamic education to the Lebanese community of Senegal and has taught them the Qur'an, ***hadith***, and ***fiqh***. He likens his daily activities to that of a judge: community members come to him for religious advice and he mediates many disputes, even between Lebanese in Senegal and family members in Lebanon. If community members want to get married, have problems with their spouses and want a divorce, have disputes over inheritance rights, or business deals that involve buying, selling, hiring, and dividing up shares, they must make an appointment with the sheikh and discuss the details of these affairs with him. He also deals with issues such as sending children to school and choosing a doctor for medical treatment. He has worked to gain Lebanese nationality for the *métis* (children of Lebanese/Senegalese unions). Even Lebanese Christians go to the sheikh for advice.

The sheikh also considers political concerns of the Lebanese community. Senegal won independence from France in 1960, but as ties with France have improved, the Senegalese began to view Lebanese (and Mauritanians) as the cause of many of Senegal's ills. They not only consider the Lebanese to be white, but also to be *Naar*, which means Arab in Wolof, which the Lebanese perceive as a racist classification. Differing views of race often result in incidents of violence, which become more frequent during key political moments. During Senegal's elections in 2000, Abdul Monem, along with the Maronite Catholic priest, met with the Lebanese community and instructed them not to react to provocations before the elections. When the Lebanese were forced to leave Liberia in the 1990s, Sheikh El-Zein organized the community in Senegal to raise money to aid them.

Such issues are not always so clear-cut. Lebanese marriages are relationships that take place between two families, not between two individuals, and difficulties often result from that. Furthermore, a Muslim woman can wed only another Muslim. If a Muslim desires to marry a non-Muslim at the Islamic Institute, the non-Muslim must first convert to Islam. Abdul Monem has conducted approximately 200 Muslim-Christian marriages, in which the bride is usually a Lebanese Muslim and the groom is a Senegalese, European, or Lebanese Christian who has converted to Islam. In Islam, the child inherits his/her father's religion, so Islam is especially strict about Muslim women marrying Muslims. Divorce is discouraged in Islam, and when community members come to Abdul Monem for advice on dissolving their marriage, he counsels them first. He tries to encourage the family to work out their problems, requiring the couple to continue to live together for a few months to be absolutely certain that divorce is the only solution.

Sheikh El-Zein encourages the Lebanese community to follow the five pillars of Islam: the profession of faith, prayer, almsgiving, fasting during the month of Ramadan, and pilgrimage to Mecca. Every year he organizes a trip from Dakar to Lebanon and then to Mecca, arranged by a Lebanese travel service that provides transportation, lodging, and food. The trip lasts two weeks and includes thirty to fifty men and women. Beforehand, Abdul Monem conducts a class on how to perform hajj, simulating the Ka'ba, the sacred building at Mecca, out of cardboard, practicing the rituals of the pilgrimage, and requiring a few volunteers to pray in front of the replicated Ka'ba so he can correct any mistakes in advance.

A sheikh is not only a teacher, a counselor, and a judge, but also a scholar. Abdul Monem has written several books on Islam, which he sells to the community. His books are written in Arabic and translated or transliterated into French for the many Lebanese in Senegal who are not literate in the Arabic script. Subjects range from a collection of prayers to two volumes on the family of the Prophet, a book on Islamic law and doctrine, more specialized books on the meaning of the Ramadan fast, and the history of Karbala where the Imam Husayn was martyred.

The Islamic Social Institute: The Sheikh's Second Major Accomplishment

At first Abdul Monem worked from his home, and it took him years to learn the ways of Senegal, to establish himself among the local religious leaders, and to convince the Lebanese community to pay their religious taxes so he could carry out his work. In 1973, he was finally able to buy the piece of land that would become the site for the *Institution Islamique Sociale* in 1978. The Islamic institute is a multipurpose organization, which has been granted NGO status. The Sheikh stresses that the institute is an *Islamic* institute, not a *Shi'ite* institute and that the Arabic language and the Qur'an are neither Sunni nor Shi'ite, but Muslim. Although Abdul Monem is a Shi'ite sheikh, the vice president of the institute is a Sunni Muslim, and the sheikh addresses the Sunnis in the community as "our Sunni brothers." Many Sunni Muslims attend Friday prayer at the Lebanese mosque because Abdul Monem is an influential man or to hear his *khotba* (Friday sermon). Also, the Islamic Institute is the only mosque in Dakar where the Friday sermon is conducted in Arabic; imams at the other mosques preach mostly in Wolof. There are Senegalese, Moroccans, Mauritanians, Algerians, and Lebanese Sunnis, among others, who attend Friday prayer with the Lebanese Shi'ite community. The only exclusively Shi'ite activities conducted at the institute are the *ta'ziyas,* or recitals of the sufferings and martyrdom of Husayn, grandson of the Prophet Muhammad, during the first ten days of the Islamic lunar month of Muharram, a mourning period known as *'Ashura*.

The Institute has a large formal lecture hall on the ground floor in black and white tile, with separate entrances for men and women. On the first floor is the mosque, a rectangular room with green carpeting and decorated with a motif of red arches and geometric designs. There is a rather small room behind the larger room for women to pray, and it has a door that is opened during prayer and closed afterward. Not many women attend Friday prayer, which is not considered a sacrament for women as it is for men. During major holidays when more women come to the mosque to pray, the larger room is divided into sections for men and women.

Administrative offices are located on the first and second floors, and the third floor houses Abdul Monem's office. It is lined with bookcases filled with books: collections of Shi'ite and Sunni law, *tafsir* (commentary) on the Qur'an, Islamic history, *tarjamat* (traditional Islamic biographies) of kings and other rulers, books on Christianity and Judaism, and books that he has published. Adjacent to his office there is a comfortable air-conditioned room with plush chairs and sofas, green furniture and carpeting, a red rug, chandeliers, and black Islamic decorations on the wall. There he holds religious classes for men and women. Attached to the building is a clinic that provides free medical treatment to the poor and which also helps the needy to travel to France, Europe, or Lebanon for specialized medical treatment that is not available in Senegal. Approximately 10 percent of the patients treated at the clinic are Lebanese and 90 percent are Africans. The building also has a kitchen and guest rooms, and the sheikh and his family reside in the upper floors of the institute.

In addition to the Islamic institute, Abdul Monem has established a primary and secondary school in Dakar that offers courses on Islam and the Arabic language, in addition to the regular Senegalese curriculum. The school is patronized by a mix of religious sects: children from different Senegalese Sunni Muslim brotherhoods, such as Tijan and Murid, Shi'ites, and even some Christians attend the "Sheikh's school," as it is called. The student population is approximately 20 percent Lebanese and 80 percent African. More recently, he founded the *Institut de Langue Arabe* that offers instruction in the Arabic language for both children and adults, and computer literacy classes taught in both French and Arabic.

'Ashura: Women's Participation in the Shi'ite Holiday of Mourning

During the ten days of *'Ashura*, I attended the afternoon *ta'ziyas* for women. The lecture hall of the Islamic institute was filled with a crowd of three to four hundred women and a few children dressed all in black, the color of mourning. Outfits varied from the traditional Muslim women's *'abaya*, a loosely fitting long-sleeved gown, to tight-fitting pants or skirts and sweaters. The room was decorated with black wall-hangings with colorful pictures and Arabic script illustrating the battle of Karbala and the martyrdom of Husayn, who was killed along with his family. A female *khatiba* (preacher) is brought annually from Lebanon to recount the story of the death of the Imam Husayn. Women came around with boxes of pastries and bags of doughnuts and biscuits flavored with anise and sesame seeds, and they also passed around boxes of tissues to those who were eating and those who were crying. The audience was moved by the recitals, which at times became quite emotional as the *khatiba* began to wail in sorrow, and the women cried, some passionately, dabbing their eyes, red with tears, and convulsing in sobs. At various points during the recitations, they beat their hearts in rhythm to the song. At one point an older woman in front of me started to suffocate from crying. The women around me sat her down, pulled off her veil, took off her glasses, splashed water on her face, put perfume under her nose, and made her drink. After a few minutes, she quickly put her veil back on and stood up again in prayer.

The holiday of mourning is even more meaningful and difficult for Shi'ites today because the events that took place in Iraq throughout Shi'ite history parallel today's politics. In 2003, *'Ashura* began only a few weeks before George W. Bush declared war on Iraq. Lebanese women were brought to tears thinking about the continuing injustice in the world and the suffering of the innocent. When the *khatiba* finished the day's segment of the story, the Sheikh came in and lectured on various topics. Baskets were passed around to collect charity donations. After each *ta'ziya*, women filed out of the lecture hall greeting each other, and they were surrounded by Senegalese beggars and the disabled in the street.

In Iran, *'Ashura* is marked by passion plays that reenact the story of Husayn, with men beating themselves with heavy chains and self-flagellating in mourning for the martyrdom of Husayn. Such activities are not carried out by Lebanese in Senegal. Although not a passion play, the recitation of the story becomes an emotional event for everyone involved. This is *the* Shi'ite holiday, the one that highlights their victimization in much of the world and makes today's members aware of their history of suffering. It is a solemn yet powerful event in which women conform to the ritual, its story, and history. As the largest event of the year bringing together Lebanese Shi'ite women, the occasion was used not only to teach them about *'Ashura*, but also to educate them on Islam. They learn about courage, to stand up against odds, to correct the wrongs of others vis-à-vis Islam, and to set a good example. Education, obligatory for all Muslims, is stressed as the key in the fight against ignorance. The role of women is emphasized as different from that of men in that women are the ones who raise and educate their children. The Sheikh also encourages women to become more active in the community, attend his Saturday afternoon Qur'an classes, go to Friday prayer, and volunteer for Al Hoda, the women's charity organization, especially since many of them do not work and have free time.

Friday Prayer

An example of a Friday sermon illustrates Sheikh El-Zein's style. I have chosen to discuss the sermon given on April 11, 2003, during the war on Iraq. The Sheikh begins, after the usual Islamic supplications, as follows:

> Oppressive regimes meet their end, even if this be at the hands of other oppressors in Baghdad on the same day the martyrdom occurred of the Eminent Imam, the Pure Martyr, Master Muhammad Baqir al-Sadr (may God have mercy on him). The issue of assassinating our scholars, whether by the current oppressive regime or by another regime in this world, has been an ongoing issue from times of old.

Abdul Monem continues his sermon explaining that Muhammad Baqir al-Sadr was a great Shi'ite scholar in Najaf and talks at length about his intellectual accomplishments. He highlights one of the many crimes of Saddam Hussein's administration, that of imprisoning Muhammad Baqir al-Sadr for protesting against their tampering with his book *Our Philosophy* by omitting its Islamic ideology and thus turning it into the philosophical basis for the Ba'ath party. Al-Sadr was believed to have been shot by Saddam Hussein in 1980 for not agreeing to issue a *fatwa* (religious decree) declaring that the Arab Socialist Ba'ath Party was a legitimate Islamic party.

Abdul Monem then compares this event to one in the Qur'an when Moses and his brother Aaron come before the Pharaoh and ask him not to require the Egyptian people to worship him since he is a human being and not a god. The Pharaoh could have killed the two men, but instead he summons his advisors who agree to call an assembly of magicians to test the truth of Moses and Aaron's claims. When they find that Moses' stick is indeed touched by a miracle from God, they prostrate themselves and declare their faith in God. Moses and Aaron lived. Such a comparison makes the murder that Saddam Hussein committed even more barbaric as he did not follow the precedent in the Qur'an.

Abdul Monem continues the sermon, detailing his personal relationship with Muhammad Baqir al-Sadr, which is that of master-disciple, and talks of his teacher's grand ideas of building an Islamic university to reform the *hawzas* (the traditional study circles of Iraq) and to modernize Islam.

> We live today in an era that is both amazing and remarkable; it is a highly advanced era. It is not an era of backwardness so that we are satisfied with giving and receiving instruction on matters relating only to ritual cleanliness and uncleanliness or valid and invalid business transactions. No. We are now wrestling with philosophies, wrestling with highly advanced civilizations and with formidable scientific progress. We cannot present Islam in the simple old form. Of course this does not mean we are saying that we should bring a new Islam. No, no. Islam remains as it is, but we must understand it in the new style.... Islam is not a religion of ritual worship only, of mosques only, of pilgrimage only, or fasting only. Islam is a religion, a way of life, a system that is complete and comprehensive, beginning at the mosque and ending at the manufacturing plant. This is Islam.

The Sheikh ends his sermon with a final idea, tying his message to Lebanon. He recalls the loss of another Shi'ite scholar, Musa al-Sadr, Mohammad Baqir al-Sadr's paternal cousin, who disappeared in Libya in 1978. Musa al-Sadr fought for the poor people of Lebanon and his "Movement for the Deprived" included Shi'ites, Sunnis, Druze, and Christians as its members and in its leadership; all were people of Lebanon and represented all of Lebanon.

He concludes his sermon with a prayer asking God to alter this pitiful current condition and to compensate Muslims for the loss of illustrious and great scholars by giving them other scholars in the future, so that, by the will of God, they could realize all the hopes that were lost with the assassination of these scholars.

The sermon is a very timely and powerful one. Abdul Monem packs a lot into half an hour of lecture: his stance against America and against Saddam Hussein as oppressors, his reverence for his teacher and for intellectuals in general, a lesson from the Qur'an, pride that the Lebanese have a scholar who is equal in greatness to the scholars of Najaf, his belief that Islam should be reformed and made applicable to the modern day and age and that there should be no discrimination against anybody, especially on religious grounds. All these ideas were developed through remembering the dead, their accomplishments, and why they died.

Relations With Senegal and Senegalese Muslims: The Sheikh's Third Major Accomplishment

Originally Abdul Monem came to Senegal to serve the Lebanese community, but eventually his work spread to benefit the Senegalese Muslims as well. Although the Islamic Institute and its affiliated institutions in Dakar cater to both Lebanese and Senegalese, Abdul Monem has founded five mosques and approximately 120 *madrasas* (Islamic schools) located outside Dakar and led by Senegalese religious men whom he had trained. The *madaris* range from larger schools to simple one-room learning facilities and teach the Arabic language and the Qur'an to Senegalese villagers.

In addition, the Sheikh visits Senegalese religious organizations, and their leaders visit him in return. With a small delegation of influential Lebanese, he attends annual Senegalese religious events, such as the *magal* of Touba, the Murid festival that attracts pilgrims from all over the world, and the *gamou* of Tivaouane, the celebration of the birth of the Prophet Muhammed in the Tijan center. He also attends special occasions with a larger Lebanese delegation, such as the funeral of Abdoul Aziz Sy, the respected Tijan caliph (head of the Sufi brotherhood) and close friend of Abdul Monem.

Global Islam

Sheikh El-Zein's religious activities and influence have spread beyond Senegal. He has strong ties to Lebanon, most famously exemplified by his role in negotiating the French hostage crisis during the 1980s. Abdou Diouf, former president of Senegal, personally called on Abdul Monem to help free the French hostages taken by Hizbollah in Lebanon. Although the conflict did not concern Senegal, Abdul Monem agreed. From April 1987 to May 1988 he traveled between Paris, Beirut, and Tehran conveying and discussing conditions for the release of the hostages, such as freeing Lebanese and Iranian prisoners in France, and he personally saw to the release of the hostages.

Abdul Monem has also attended numerous international conferences and dialogues on Islam and has published his papers and speeches in a book printed in Lebanon. He has presented papers at a conference in London connecting Shi'ite Islam to World Islam, at a conference in Tehran on Islamic Thought, and at a conference on the Holy Qur'an in Qom. He spoke during the tri-annual Islamic Summit Conference of the Organization of the Islamic Conference during its 1991 meeting in Dakar, which was the only time the international Muslim organization had convened in sub-Saharan Africa. (It will be held again in Dakar in 2008). He also met with Pope John Paul II during his visit to Dakar in 1992, where they discussed the position of Christianity vis-à-vis Islam and the Israeli occupation of southern Lebanon. Sheikh El-Zein's activism did not end there. He wrote a letter to the United Nations Security Council deploring the situation in Iraq under Saddam Hussein, and he also participated in a Muslim-Christian dialogue in Detroit.

Sheikh Abdul Monem El-Zein is known throughout Senegal and Lebanon, as well as other parts of the world, because his charisma leaves a lasting impression. Community members attend his late-night Ramadan lectures and other holiday celebrations because Sheikh El-Zein can inspire and hold his audience with powerful messages, impassioned tones, and dramatic pauses. His patience is enduring: one Lebanese woman spends hours with him every week so that she can learn proper Muslim values from him to back up her decision to start wearing the veil. Some Lebanese shopkeepers revere him enough to display his picture in their stores, alongside the pictures of other Lebanese sheikhs such as Musa al-Sadr, and they also display the pictures of one or two Senegalese clerics to please the local population and encourage their business.

When I was deciding whether or not to attend his Ramadan lectures in November 2002, despite a multitude of phone calls from community members informing Abdul Monem of a possible spy in the audience, he told me a parable of Juha, the Arab fool. Juha and his father were riding a donkey, but Juha felt sorry for the donkey for having to carry too heavy a load, so he asked his father to step down. People watched Juha riding the donkey while his elderly father walked, and they criticized him, so he too got off the donkey and walked next to his father, leaving the fortunate ass with no load at all. The moral of the story is that no matter what you do, whether you ride or walk, or decide to be present at his lectures or remain absent, people will talk. Abdul Monem disclosed to me that he wanted the world to understand the truth about who he is and what he does, so he decided to work with me in order to counter Western misconceptions that all

Muslims are terrorists. The bright side of today's politics, Sheikh El-Zein informs me, is that with the Iranian revolution now in the background, and al-Qaeda, bin Laden, and Saddam Hussein seen as evil Sunni Muslims in the public mind, converts are more and more choosing Shi'ite Islam.

CHAPTER 7

Son of an Ayatollah: Majid al-Khu'i *(Iraqi Religious Leader in Great Britain)*

John and Linda Walbridge

Sayyid Majid al-Khu'i asked his mother, "Why, if I work and study, don't I get a salary like everybody else?" After all, his father was the Grand Ayatollah Abu'l-Qasim al-Khu'i, the most important Shi'ite cleric in the holy city of Najaf in Iraq and probably the most distinguished Shi'ite cleric in the world. His mother agreed to intervene on his behalf with his father. The next day Ayatollah al-Khu'i said to his son, "You know how to get a salary. You go to my office and take an exam like any other student. If you pass it, you get a salary. Why don't you just go take this exam?" Majid went to the office where the exams were administered and told the ulama, or clerics, in charge that he wanted to take the exam that would make him officially one of his father's students. "But you don't need to take an exam," one of the ulama responded. "I know," said Majid, "but I want a salary, and I can't have one unless I have passed the exam." During the exam, Majid became worried and started to have doubts about whether he really needed this salary after all. How about if he did not do well and embarrassed his father before the other ulama? His fears seem to have been unfounded. His father called him to his presence that night and said, "Very good. I am happy. Now you can have your salary."

However, after several months, he said to his father, "Listen, Daddy, you give me the same salary that is normally given to students. But there is a difference between me and the others. First, when I go to visit the holy shrines, so everyone on the bus looks to me to pay since I am the son of the Grand Ayatollah. But one bus ride and my salary is gone! And it is not fair for me to keep my hand closed and wait for someone else to pay. I don't like that. Second, other students also have salaries from other ulama, but I don't have that. I just have a salary from you. It is not enough." His father replied, "There is another difference as well. When I give to other students, I give two Iraqi dinars and they have to live on this. But you live at home, and your food and clothes are free and your books are free. The two dinars are just your pocket money."

There was a difference in age of about sixty-five years between the Grand Ayatollah and his son. Majid tried to convince his father that times had changed and that with the oil money flowing into Iraq, many people were prosperous. But his father insisted that a Grand Ayatollah must live more humbly than the rest of the people. After all, he asked, when a poor peasant came to Najaf to pay the religious tax to him as the representative of the Hidden Imam, was it fair for him to have luxuries such as a car, a refrigerator, or a television? It was not proper for him to waste the Imam's money on luxuries for him and his family.

Ayatollah al-Khu'i finally did accept a car as a gift from some wealthy Kuwaiti Shi'ite. The car enabled him to commute the five kilometers from Kufa, where the family lived, to the mosque and his offices in Najaf more efficiently than the two-horse carriage that he was accustomed to traveling in. However, money always remained an issue. For example, he would turn off the lights as he left a room so as "not to waste the Imam's money." When his car needed repairs one day, he and Majid were forced to travel by taxi. After prayers at the mosque in Najaf, Majid went to find the driver. He told the driver that the people liked to talk to his father after prayers, and so he might have to wait. The driver replied that he should be paid a hundred fils more than the normal half dinar if he had to wait. Majid agreed. However, on the way home, his father asked him about the fare for the cab. When Majid told him that he was paying six hundred fils, his father chided him, telling him that he was paying more than the going price for a cab ride from Najaf to Kufa. Majid wanted to know how he knew the fare for the cab ride. His father replied, "I need to know how the people live so I ask them about such things."

Linda met Majid al-Khu'i in the mid-1990s, perhaps fifteen years after he had taken his exam. She was working on a book on religious leadership among the Shi'a and had come to London to visit the Khoei Foundation, an influential Shi'ite charity founded by Ayatollah al-Khu'i, who had died in 1992. She had already met Yusuf al-Khu'i, a polished cleric in his middle years who was the grandson of Ayatollah al-Khu'i and who was the assistant director of the foundation. A handsome man in his early thirties walked in. He was dressed as a cleric and introduced himself. "I'm Majid al-Khu'i," he said. "But you're too young!" she blurted out. He laughed delightedly. He explained that he was the son of Ayatollah al-Khu'i's much younger second wife, while Yusuf, his nephew who is considerably older than him, was the grandson of the Ayatollah and his first wife. Though Majid now lived in London and spoke excellent English, his position was shaped by the theological politics of Shi'ism and its holy cities and ultimately by the events that took place in Iraq more than thirteen centuries ago.

The Imamate and the *Marja'*

Majid's home city of Najaf, about a hundred miles south of Baghdad, is the site of the tomb of 'Ali, the Prophet Muhammad's nephew and son-in-law. According to the Shi'a, 'Ali was the first of twelve Imams, the appointed successors of the Prophet. The Imamate is thus almost as old as Islam itself. When the Prophet Muhammad died, the leadership of the Muslim community followed the traditional democratic Arab custom: the community at large formed a consensus that the most senior, respected follower of the Prophet would assume the role as *khalifa* (caliph) of this young and expanding religious community. But there were those who argued that the cousin and son-in-law of the Prophet, 'Ali ibn Abi Talib, should assume that role since they were convinced that this is what the Prophet had expressly wished. The dispute rent the early Islamic community. 'Ali eventually became the caliph two decades later, only to be murdered by an extremist. Two decades after his father's death, 'Ali's son Husayn revolted against the caliph Yazid, the corrupt son of 'Ali's old enemy. He and his handful of supporters were killed at Karbala in Iraq. The Shi'ites mourn him still during the grim holy days of 'Ashura, the first ten days of the Muslim year. According to the largest surviving Shi'ite group, the line of rightful heirs and descendants of Muhammad and 'Ali continued down to the eleventh generation. The twelfth Imam vanished from sight, though he is believed to be still alive and occasionally appears in the flesh or in dreams to his faithful followers.

Majid's father occupied a position defined by the claims of the Hidden Imam and the complications of Islamic law. In Shi'ism, as in the other sects of Islam, law occupies a central place. Every conceivable human action is either obligatory or forbidden, desirable or undesirable, or indifferent. For Shi'ites, the details of this law are to be deduced from the recorded words and actions of the Prophet and the Imams. The believer is expected to make an effort in good faith to determine what the Islamic law is in any given circumstance and to follow it to the best of his ability. There is a difficulty here, for even the greatest of scholars often differ on the details of the sacred Law, especially as it applies to new circumstances. What then is the ordinary believer to do, who must pursue his trade and cannot devote his life to legal scholarship? For modern Shi'ites the answer usually lies in finding the person whom he believes is best qualified to make independent judgments on religious law and then follow his advice. Such a person is called a *marja'* (a source of reference) or more commonly "ayatollah," a title of respect literally meaning "sign of God." At any given time, there are only a handful of grand ayatollahs in the world; these are the *marja's* with a large number of ordinary Shi'ites as followers.

If you ask a devout, observant Shi'ite how he chooses an ayatollah to follow, he will say that he selects the person most knowledgeable and pious. Such a person must first be a *mujtahid*, a cleric who has not only completed the full course of study at a madrasa (seminary) in Iraq or Iran, but is also recognized by other ulama for his learning and his ability to interpret and make legal judgments. It requires success at a course of study that is equivalent to a rigorous PhD. Among these highly trained *mujtahids*, the Shi'a are supposed to come to a consensus on who best meets the criterion to be the *marja'* for the community of believers. It is not permissible to follow a deceased *mujtahid*, so when a *marja'* dies, the process of recognizing the most learned *mujtahid* begins immediately.

There is a second difficulty. According to Shi'ite belief, the Imam has certain responsibilities and rights—appointing leaders for Friday prayers, declaring holy war, and collecting religious taxes. What is to be done about these until the return of the Hidden Imam? In practice, the most important of these is the *khums*, the religious tax comprising one fifth of what is left when living expenses are subtracted from income, half of which is owed to the Imam and half to the family of the Prophet. Shi'ites usually pay this tax to the grand ayatollah they thought most worthy. For Shi'ites outside Iran and for many in Iran, this grand ayatollah has usually been the most eminent of the clerics living in Najaf, the city that grew up around 'Ali's tomb.

The responsibility of paying and receiving this tax is not taken lightly, for it is a debt that the believer will answer for on the Day of

Judgment. It should be given only to a cleric known to be worthy. Likewise, it is a heavy responsibility for the grand ayatollahs who receive the tax—money that is received as a trust for the Hidden Imam is to be spent only in the service of Islam. If Majid had asked his father why he cared whether he paid the taxi driver an extra hundred fils, he would doubtless have been told that the money belonged to the Hidden Imam and that he was answerable for how it was spent. In any case, an ayatollah who spent this money on himself would very soon see the contributions going to some other more scrupulous ayatollah.

When Majid remarked that people were more prosperous these days, he was not exaggerating. Many Shi'ites who followed al-Khu'i had done very well. An immigrant to America who owned a couple of gas stations might owe *khums* of many tens of thousands of dollars; a wealthy businessman in the Gulf might owe millions. All of this money was to be used only for the service of Islam and Shi'ism. And there were more Shi'ites who paid this tax to Ayatollah al-Khu'i than to any other *marja'*.

For the majority of Shi'ites in the world, Majid's father, Grand Ayatollah al-Khu'i, was the perfect model of the *marja'*. He had come to Najaf as a boy of twelve in 1912 and had remained there studying and teaching throughout his long life. He taught more students than any other contemporary ayatollah. In the mornings, he read at his modest house in Kufa, 'Ali's old capital, and then he was driven the few kilometers to Najaf, where he gave lectures, discussed religious and legal issues with other clerics, conferred with the staff who tended to the administration of his offices, and responded in writing to the endless questions put to him by Shi'a from all over the world. His routine varied little, and there were constant demands on his time. When not formally engaged in work, there was an endless procession of believers at his door. They wanted to bask in the presence of a man so revered that his photograph adorned the homes, mosques, and Husayniyyas of Shi'as around the world. In these places, one could also find copies of his books, faxed copies of his legal rulings in his frail handwriting, and sometimes lists of people with the amounts that they had paid as *khums*.

His routine did not vary much from that of the great *mujtahids* of a hundred or more years ago. There was also no variation in his style of life, his appearance, or the subjects he taught. He did not value novelty and innovation. He treasured and cultivated the rigors and intricacies of Islamic law and the disciplines associated with it. His method of proving a point was the same as that used by countless generations of his predecessors: questioning others in such a manner that their ignorance would ultimately be revealed and his knowledge and wisdom made manifest. The books he wrote, all on traditional subjects, fill a long shelf. He was a professor's professor, a quiet man devoted to his books but with an air of saintliness.

Disaster and Exile

In 1978, about the time that Majid was asking for his salary, a revolution broke out in neighboring Iran. The following February a prominent Iranian cleric, Ayatollah Khomeini, returned to assume the leadership of the new Islamic Republic of Iran. Majid would have known Khomeini, who had spent most of his fourteen years of exile teaching in Najaf. Soon, Iranian propaganda broadcasts were calling for an Islamic revolution in Iraq. Majid's father avoided taking positions on political issues, following an old Shi'ite tradition of political quietism, but many of the younger clerics in Iraq sympathized with the Iranian revolution. Saddam Hussein and his secret police feared the worst and arrested large numbers of clerics, and executed many of them. There was hardly any Shi'ite clerical family in Iraq that did not lose someone; dozens of members of some families were killed. Many clerics and their families fled to Iran or to the West.

Ayatollah al-Khu'i stayed on in Iraq, protected by his prominence and his avoidance of politics. Majid stayed with him, though by this time most of the Ayatollah's affairs were being run through the Khoei Foundation in London. In 1980, Saddam invaded Iran, fearing that if he waited, Islamic Iran would grow strong and destroy his government. The war quickly became a bloody stalemate, and families that had not lost members to the secret police lost members in the war. Throughout the war the persecution of the Shi'ites, especially of the Shi'ite clerics, continued. The *hawza* (the community of teachers and scholars in Najaf and Karbala) dwindled as teachers fled or were arrested, and few students dared to come. Saddam then invaded Kuwait in 1990, hoping to find some compensation for the ruinous costs of the long war with Iran. That winter, a coalition led by the United States struck back, shattering the Iraqi army and driving it out of Kuwait. President George H. W. Bush called on the Iraqis to revolt and overthrow Saddam. Believing that their liberation was at hand and that the Americans would support them, the Shi'ites of southern Iraq and the Kurds of the north rose against Saddam's government. Majid's father attempted to moderate the passions of the revolt, issuing a legal ruling that prohibited personal acts of vengeance. Majid was a member of a committee of twenty-five that attempted to maintain order. President Bush, however, had no intention of intervening in Iraq. Using tanks that he had been able to withdraw from Kuwait and helicopters that the truce agreement permitted him to use, Saddam crushed the rebellion. Tanks stormed the holy cities, even breaking into the holy shrines to slaughter women and children who had taken refuge there. American armored units watched from the edge of the desert.

Majid fled with hundreds of thousands of other Iraqis. His father stayed behind in Najaf and died the next year under house arrest. The world had a last glimpse of him when he made a forlorn appeal for peace on Iraqi television.

London and the Khoei Foundation

A *marja'* who becomes a grand ayatollah acquires his position through his reputation for piety and learning. That reputation is spread by his sons, sons-in-law, and students. Grand ayatollahs do not travel, and they act through their representatives. Majid had already started to perform this function in Iraq, but after the assassination of his brother Taqi in 1994, he assumed the leadership of the Khoei Foundation, the institution managing the ayatollah's charitable and religious projects. The foundation had been established because of the large amounts of *khums* being received by Ayatollah al-Khu'i. The amount was so great that questions were being raised about whether it was being spent properly. The foundation conducted activities all over the world, including running a school in London. It built mosques, ran schools, and operated other charitable enterprises. These activities in turn spread the reputation of Ayatollah al-Khu'i, thus firming his position as the leading Shi'ite cleric in the world. When we lived in Dearborn, Michigan, for example, the Khoei Foundation had promised funds to build a large mosque in the community. And it was al-Khu'i's legal rulings that were posted in the mosque, whose imam proudly proclaimed himself to be a representative of al-Khu'i. After the failure of the 1991 revolt, there were large numbers of Iraqi exiles, and impoverished people still in Iraq, who had to be provided for.

The activities of the foundation continued even after al-Khu'i's death in 1992, but Majid was forced to deal with new problems. First, a Shi'ite is not supposed to follow an ayatollah who has died. Since none of al-Khu'i's sons were qualified to assume their father's position, discrete negotiations began with the clerics in Najaf to find a new *marja'* who could be endorsed as an ayatollah, in the process snubbing the efforts by the Iranian authorities to persuade the Shi'ites formerly loyal to al-Khu'i to ally themselves with an ayatollah in Iran. In the end, like most Iraqi Shi'ites, they united behind Ayatollah Sistani in Najaf, an eminent student of their father.

There were also other pressing political issues. There were Iraqi exiles of every sort in London, all of whom were convinced that Saddam would have to be overthrown, and many of them were plotting to gain power in a future regime. The interests of the Iraqi Shi'ites had to be represented on the world scene. Moreover, it was imperative that the Iraqi Shi'ites distinguish themselves from the Iranians, whose bloody Islamic Republic had acquired a very bad reputation in the West.

Majid threw himself into his new roles as philanthropist and diplomat. His schoolboy English improved rapidly. He continued to wear the clerical garb, but his personal charm won him many friends. Like many Muslim clerics who find their way to Europe or America, he thoroughly liked the West. He enjoyed its freedom and openness, and he liked the people. Now he had a family too—a young wife and a child. Under his leadership the foundation was recognized as a non-governmental organization by the United Nations and it became the most prominent Shi'ite organization in the West. Majid also guided the foundation in its dealings with other Iraqi exile groups attempting to overthrow Saddam.

Return and Death

With Saddam's disastrous defeat in Kuwait, which closely followed the bloody stalemate of the Iran-Iraq War, most observers, including the American government, expected that Saddam would be overthrown. But he was not, and he was able to reconsolidate his power despite sanctions intended to force him to give up the suspected nuclear, biological, and chemical weapons programs. As the 1990s passed, there were various efforts by the American government and American intelligence agencies to undermine or overthrow Saddam; none was successful. When the second Bush administration took office, a number of its senior members were veterans of the first Bush administration and viewed Saddam as unfinished business. After the September 11, 2001, terrorist attacks, Bush decided to overthrow Saddam.

The American government turned to Iraqi exiles as allies, advisors, and potential members of a post-Saddam government. The Shi'ites were seen as important potential allies against Saddam. Majid al-Khu'i was one of the people the Americans approached. He was eager to go back to Iraq. There were good reasons not to, of course. No one imagined that the motives of the Americans were pure, and it was known that they had abandoned local allies earlier: notably the Kurds of northern Iraq in the 1970s. Those allying themselves with the Americans were likely to be seen as traitors by other Iraqis. The political situation was volatile and extremely dangerous. And Majid had a young family. Many of his friends urged him not to go.

However, Majid thought that the needs and opportunities were so great that he had to go. The Shi'ite areas of Iraq were desolate from years of war and oppression. Many of the institutions of faith and learning in Najaf and Karbala had been closed for years. The overthrow of Saddam, now a certainty, was likely to result in an outburst of revenge and bloodletting of the sort he and his father had attempted to prevent twelve years before. At one time, the holy cities had been managed by a delicate web of relationships among the prominent clerics and families. These were now shredded by years of oppression. If nothing was done, there would be conflict and bloodshed between those who had suffered under Saddam and those who had survived by compromising with him. Majid believed that by intervening in the situation early, backed by the resources and prestige of the Khoei Foundation, the basis might be laid for the restoration of a peaceful and decent community in the Shi'ite regions of Iraq and in the holy cities. At least he had to try.

Then there was the larger political situation. If they did not act as a united force, the Shi'ites might be shut out of political power in the new Iraq, as had happened after the fall of the Ottomans eighty-five years earlier. Majid thought that the Americans offered the best chance for a good political outcome in Iraq, although he could hardly trust them. The Americans were eager to work with him, since early support by a legitimate and widely respected Shi'ite leader would gain them support from the Shi'ites, who were, after all, the majority of the Iraqi population.

Majid returned to Iraq on an American plane on April 1, 2003. The fighting was not yet over. A military helicopter flew him and his entourage to Najaf, where he was greeted with delight as well as suspicion. He and his family were still remembered in Najaf, of course, but people were uneasy about the exiles who returned. The exiles had fled to a life of safety and comfort abroad while those who stayed had suffered for another decade under Saddam. Now they were returning under the protection of a foreign army and with the support of foreign intelligence agents to assume positions of power in the future government. And Saddam might still come back, as he had in 1991, to take terrible vengeance on those who had allied themselves with his enemies.

The holy cities also had their own politics. Najaf and Karbala had always been turbulent places where clerical factions and families had competed for prestige and control of the vast revenues of the shrines. Often these rivalries had been fought out on the streets between gangs allied with the various clerical factions. The oppression of the Saddam years had only made these tensions more bitter. Those who had stayed behind and suffered resented those like Majid al-Khu'i who had survived and prospered by fleeing. Some had retained their positions by making common cause with Saddam. Now everything was open for grabs. Once, for example, there had been three locks on the offerings box at the shrine of 'Ali in Najaf—one key held by the representative of the government and the other two by prominent clerics. Now the locks had been broken and the offerings looted.

Majid immediately began trying to win support and stabilize the situation in Najaf. He distributed a third of a million dollars to the poor in Najaf—money from the foundation, he insisted, not from the Americans. Then he took the more dangerous step of trying to reconcile one of the most bitter hatreds in Najaf, that between Haydar al-Rufa'i, the custodian of the shrine of 'Ali, and the persecuted clergy. The Rufa'is had been for centuries the custodians of the shrine. Haydar al-Rufa'i had made many enemies by accepting the position under Saddam. Rufa'i had not been to the shrine since Saddam's forces had left the city. Majid went to his house in the morning on April 10 and brought him to the shrine, hoping to effect a reunion. No one knows for sure who was responsible for what happened next, but Majid's friends blame the young Muqtada al-Sadr.

Al-Sadr was the son of one of the grand ayatollahs who had stayed behind in Iraq. When al-Sadr's father had acquired too much popular following, Saddam ordered him assassinated, which caused riots in the Shi'ite areas. While Majid and his father represented the old legal and intellectual tradition of Shi'ism, Muqtada represented an even older charismatic tradition. Using the messianic symbols associated with the Hidden Imam, Muqtada al-Sadr had built up a large following in the Shi'ite slums of Baghdad. Now he was making a play for power in the chaotic aftermath of the invasion.

Majid and Haydar al-Rufa'i went to the shrine of 'Ali and prayed there, and then they went to the custodian's office. Soon a mob gathered outside, shouting slogans in support of al-Sadr and demanding that the hated Rufa'i be turned over to them. Someone smashed the window. Majid tried to speak to the crowd, but he could not make the microphone work. Someone tried to stab him. One of the clerics who had come back from America fired a pistol in the air. The little group tried to defend themselves with guns from the custodian's office. A grenade wounded Majid, blowing off two of his fingers.

After ninety minutes, Majid and his friends were out of ammunition and tried to surrender. Their hands were bound and they were taken out of the office. The crowd attacked them with knives and swords. Al-Rufa'i was killed on the spot, and Majid was gravely wounded. When they were turned away from al-Sadr's house, they took refuge in a shop. Though the shopkeeper tried to convince the crowd that Majid was already dead, the crowd dragged him out and shot him. His body was dragged through the streets behind a car.

Majid's death was news for a day or two, but it was quickly overshadowed by the fall of Baghdad. Events confirmed Majid's fears. A frenzy of looting crippled the public institutions that had survived the American bombing. Though Ayatollah Sistani, the greatest of the clerics still in Najaf, tried to moderate the situation, Muqtada al-Sadr's supporters revolted against the American occupiers. The uprising culminated in the siege of Najaf itself by American troops. Much of the historic old city was flattened, and the storming of the shrine was prevented only by the direct intervention of Ayatollah Sistani. Thousands of Shi'ites were killed in these actions and also by Sunni terrorists.

To Majid's friends he is a martyr like his famous ancestors Imam Hasan and Imam Husayn. Perhaps, if peace and democracy are established in Iraq and if the Shi'ites are finally able to get their share in ruling Iraq, Majid will be remembered as one of those who died to make it possible.

PART III

Reconciled

CHAPTER 8

The Irrepressible Salma
(Afghan Aid Agency Manager)

Patricia A. Omidian

She walked into my friend's house in Peshawar, Pakistan, a bit shy of her English, but very determined to carry out her position in her agency. Small-boned, slim, with wavy dark brown hair and engaging brown eyes, Salma seemed self-contained and sure of herself, although she later told me that she had been somewhat nervous at our first meeting. She was then the gender coordinator for a large Afghan nongovernmental organization (NGO) and was nearly thirty years of age. This was the year 2000 and Salma was an Afghan refugee herself, living in the urban area of Peshawar, Pakistan. Though her agency was based in Peshawar, it provided development and emergency services to Afghans in Taliban-run Afghanistan. Like the other Afghan women I worked with, Salma was self-assured and committed to Islam, and to women's rights within an Islamic context.

At that time, I was an independent consultant based in Islamabad, Pakistan. I had earlier worked on mental health and resettlement issues of Afghan refugees in California for over ten years. This led to my research, as a senior Fulbright Fellow, on the mental health of Afghan refugees in villages in Pakistan, which has led in turn to my staying on for six years as a consultant in the region. I had received a message from a friend stating that a certain agency needed a trainer for a program that would help seven Afghan NGOs develop gender policies for their agencies. My first call for the job came from a man who was in Islamabad for some meetings. He explained that Salma was out of the country for a training workshop and that he would meet me on her behalf. I had thought that this man was her employee, but it turned out that he was her boss. We discussed the program and what was needed. He arranged for me to meet Salma the following week in Peshawar at my friend's house. In Pakistan, it is as common to conduct business at home as it is to conduct it at an office. Often, because offices are so crowded, many internationals prefer to have meetings in private residences.

The second time I met Salma, she brought out a photo album of her travels from field office to field office within Afghanistan during the Taliban era, which was a period when the movement of women was sorely restricted. She traveled at that time with other women staff. In the pictures all the women had their burqas thrown back to show their faces. They would put the burqa back on when there was any indication of a road check or block by the Taliban. Salma was very proud of her work for the women of Afghanistan and resented the Western image of the Afghan women as victims. She then felt, and still feels, that Afghan women have great needs. But she also feels that women must be the active agents in improving their own lives and livelihoods. She feels that change has to come for women, but that the change must come within the context of their own culture and society. And Salma is working hard to bring about those changes.

In Pakistan: The Life of Afghan Refugees in Peshawar

My first impressions of Salma were of a distant, intelligent, and strong young woman. But in a very short time, Salma and I became friends. She shared her life with me, inviting me into her home like a sister. Like many other refugees from Afghanistan who had fled the war and destruction in their homeland, Salma had left her country in 1992 when the *mujahidin* (the militia groups that had fought against the Russians) took over Kabul. Before coming to Pakistan as a refugee, Salma, like so many other Afghans, had lived through the many years of war and fighting that had plagued Afghanistan, since 1979 when the Russians invaded in order to prop up the Communist government in Kabul, the capital city. She and her family lived in Kabul and experienced fighting in their own neighborhood. When the fighting started between *mujahidin* factions in 1989, she and her family were in their home. She remembered hundreds of rockets hitting the city every day.

Most refugees who are registered by UNHCR (United Nations High Commission for Refugees) live in camps. These are the places that one thinks of when one hears the word "refugee." But in Pakistan, because there were so many Afghan refugees, the refugee camps became villages where Afghans could freely enter and leave. Most men worked as laborers in surrounding towns or, sometimes, even abroad. Women and families stayed put in the adobe mud-brick homes. Most camps had schools and health clinics run by international nongovernmental agencies. Pakistan was the home to approximately three and a half million Afghan refugees, and most lived in these refugee villages. In addition, there were also thousands of unregistered refugees living in the urban areas of Pakistan. Officially, these Afghans were not counted as refugees, nor were they eligible for refugee services such as free schools and low-cost health clinics.

Like most middle-class refugees, Salma lived in a neighborhood in Peshawar that had broad avenues with large expensive mansions. But the Afghan refugee families lived along small alleyways in two-story houses that had been subdivided many times to accommodate all of them. In this area it was common for a family of eight people to share two rooms. But Salma and her family had more space to live in, and they shared a house with just one other refugee family—her sister's fiancé's brother, his wife, and child. The in-laws lived downstairs and Salma's family lived upstairs. The in-laws had to live downstairs as both the husband and wife suffered from spinal injuries because of the war and both were wheelchair bound.

Because Salma was single, she lived with her mother, father, and sister; her sister was engaged to an Afghan living in Holland. Unlike in the West, an Afghan living here would not dream of having a separate apartment to live alone and away from other family members. Living alone is not desired, for it leaves one vulnerable to gossip, which damages the reputation of both the individual and his or her family.

Gossip exerts a powerful social control. Fear of gossip prevents women from venturing too far into the public realm, an area controlled by men and by public opinion. In Pakistan, middle-class Afghan refugee women found their lives more restricted than they had been in urban areas of Afghanistan, both before the war and during the Communist regime that controlled Kabul until 1992. Living among strangers in a new country gave an even greater power to gossip as a means of social control. In addition, women who join the public work force and who want to move up in status in their offices are subject to the worst kind of backbiting and gossip. This is not easy to deal with, especially if the women are unmarried. Gossip against a woman is often accepted as true, and it often results in the woman leaving her position in the office and retreating to her home. Again, it is the reputation of her family that is also at stake here.

Before coming to Pakistan, Salma had been a student at the university in Kabul. But when she and her family first arrived in Peshawar, she started a tailoring business, making clothes for neighboring women to help her family cover expenses. Because they chose to live in an urban area, they were not eligible for refugee aid, and inflation made living in Peshawar expensive.

At this time, Salma also attended computer and English courses at a local private school. After she received her certificates, she took up a job as a computer operator in a French NGO. She was at that time engaged to a man in Peshawar who was trying to leave the region and settle in Europe. Salma did not agree to this move and her fiancé left without her. Without the support of her family, she would have had fewer options and no personal choice in such a situation. In Afghan culture, a woman who is engaged is like a married woman. She cannot easily change her mind. Salma was engaged but had decided that she did not want to marry the man. She had many reasons for this decision, including the fact that she was committed to working for her country and fellow Afghans, and that she did not want to go abroad. There were also issues of personal compatibility.

Her family agreed to her breaking the engagement and staying single. This meant that she would continue to be dependent on her family and will have problems in the future if she never marries. But she was sure that this was what she wanted. In Afghan Muslim culture, it is considered that a girl has not broken any rules if her family supports her. Her father and her elder brothers encouraged her to pursue her career and stay in the region; they had no objections as long as she stayed within Afghan social bounds in attitude, demeanor, and behavior.

After this, Salma joined an Afghan NGO to work as a clerk. The management quickly noticed her skill in managing problems and people, and they promoted her to more responsible positions. But there were also obstacles, most prominently gossip. There was talk in the office that she was moving up because she was cute and the men liked her. In the West, such talk would be harmless, or only annoying at most. Here in this Muslim region, it can be deadly. Many girls (unmarried women) will be forced to stop working if their family hears that such things are being said about their daughters. If the gossip is believed, the girl could suffer beatings or worse. But Salma's family only encouraged her to be strong and brave. She stood up to the gossip by dressing in conservative but stylish clothing. At one point she even began to wear glasses to make her look more serious and older. She wanted people to take her seriously and not see her as a young woman looking for a husband.

By the time I met Salma, she was the gender advisor for her agency and was running the personnel department. She was in charge of recruitment and took the job seriously. She supported the hiring of women and girls in every sector of the agency. Her goal was to have at least 50 percent women staff in the agency. She worked with an aim to teach the men in the agency, many of whom were very conservative, that Islam allows women to work to support themselves and their families. She highlighted issues of gender discrimination and acted quickly to address those problems. The agency had a rule that all the staff had to attend a three-day gender awareness course. Salma conducted the training. Because of her efforts, this agency became known as one of the most gender aware and gender balanced agencies working in Afghanistan. And all this was achieved during the time of the Taliban, when women were denied the right to work by the Taliban government.

By following society's rules and the rules of Islam, Salma sought to prove to men and women around her that there was no discord between working and adhering to Islam's tenets. She even traveled to field offices in Afghanistan wearing the burqa, the covering that the Taliban required all women to wear. Salma thought that it was a nuisance, but she said that as long as she could work she did not mind wearing it.

Back to Afghanistan: Life in Kabul After the Taliban

In September 2001, life changed again for the Afghans and for Afghanistan. The United States was hit by the worst attack on American soil in its recent history, and in return it struck at Afghanistan's Taliban regime. I know of no Afghan in Afghanistan, including those I met and worked with in Kandahar, home of the Taliban, who did not welcome this action. But because I was evacuated out of Pakistan to the United States on September 14, 2001, I could only keep contact with Salma through e-mail. She and her other colleagues would send me messages to tell Amerians that the Afghans needed their help and wanted the Taliban removed. They all wanted to return to Afghanistan and work to rebuild their country.

The United States started bombing Afghanistan, particularly Kabul, in October 2001. By the end of November, the Taliban had fled from Kabul and were hiding in the mountains to the east along the border with Pakistan. They are there to this day. With them went the other non-Afghan fighters, including Arabs and Chechens. The

American policy of using air power was welcomed in Kabul, and there are many legends about the accuracy of their weapons. However, rural Afghans have a different view. They wanted the Taliban out but did not want their culture violated. The war there is not over yet.

By December 2001, I had returned to the region. This time my colleagues and I were in Kabul, living in the office because there was no housing. Kabul had almost been leveled in the violence of the past twenty-five years, particularly after the Russians pulled out of the country in 1989, leaving the antigovernment groups, the *mujahidin*, to fight over who would control the capital. In this fighting, the capital city of almost a million people had been largely reduced to rubble and minefields. When we returned to Kabul in 2001, over 65 percent of the city was destroyed, electricity was barely available, and water in the wells was contaminated. Our only option was to live in the office.

Salma and four other Afghan women staff members lived in one section of the office, while the men stayed in a separate building. Fear of gossip kept the women from leaving the compound except for work. This was a difficult time for all of us, but Salma worked hard to ensure that the staff was aware of the needs of their female colleagues. She also hired many local women so that there would be enough women in the office to counter any gossip that might spread. As always, it was gossip that controlled women's actions and movement and prevented most women from venturing far from home.

These courageous women worked hard and were able to counter the gossip by their actions and their carefully circumscribed behavior. They effected the transition of a large NGO, from the status of functioning in exile to one centered in their own country. These women were largely responsible for moving the office to Kabul, for the continuation of education, health and development programs, and for the return of many families connected to the NGO. Salma was tireless in her efforts to support these women and to ensure that men, both in and out of the agency, would not talk about the women or gossip about them. Salma had no doubts that any white collar job that is usually done by a man could be done by a woman. This was rare in post-Taliban Afghanistan.

As more Afghans returned to Kabul, Salma was able to bring her family, her mother and younger brother, to Kabul. They reclaimed their father's home in the western section of Kabul. This was once an affluent area of Kabul but had been completely destroyed by the rockets of the *mujahidin*. Salma's brothers started rebuilding the home to create a place where memories of her father and old times could be rekindled. It was a labor of love that Salma put her heart into.

Then tragedy struck. One night a gang of armed robbers entered the home held the family at gun- and knifepoint and stole everything. Everything! They held Salma with a knife and tried to force her mother to tell them where the money was hidden. In their search of the house they found very little money or gold jewelry because Salma had spent most of her family's savings on rebuilding the house. In the end they stabbed Salma in the shoulder and escaped. After this the family did not feel safe anymore and their hope of restoring their home was shattered.

Then they rented an apartment in Mekroyan, an old Russian-built housing area near the Kabul airport. Her other brothers also brought their families to Kabul. For a while they all lived together in a small apartment because housing was still scarce. None of the returning staff lived as well as they had done as refugees in Pakistan. The stress was enormous as the apartment's water system was damaged, there was no electricity, and the buildings were cold. Further, the house was crowded with nine people sharing five rooms. Although her brothers lived with her, it was Salma who was in charge of her mother and younger brother. Eventually her other brothers moved out to their own homes, easing the level of stress she was under.

It was during this time that Salma acted to make one of her own dreams come true. She wanted to adopt a son. In Afghan society there is no mechanism for adoption. A child can be cared for by another family but will never be considered a true family member. An adopted child does not inherit from his or her adopted parents, nor do the adopted parents have any kind of permanent claim over the child. Should the parents of the adopted child want him or her back at any time, the adopting family is obligated to return the child. One example for which I have personal experience is that of the wheelchair-bound couple, relatives of Salma, who lived in Peshawar with her. They had an adopted daughter who was actually the child of one of the man's brothers. When he died and his widow had to return to her brother's home, this daughter, whom they had raised from infancy and who considered them as her parents, was taken back by her birth father. She will be lucky if she ever sees her adopted mother again, because they live in different parts of the country.

Salma, as a single woman in a traditional society, needed to plan for her future. She wanted to stay single and did not want to be dependent on her brothers. At the same time, she wanted to have some security in her old age. Afghanistan does not have a social security system for the elderly—that need is catered for within the family. For

Salma to remain independent, she needed a son. Her only option was to try and adopt one.

At that time in Herat, a city far to the west of Kabul, a woman gave birth to twin boys. The woman was very poor and already had many children. She lived in an internally displaced persons camp near the city. After she gave birth to the babies, one of them died, and she handed over the other baby to Salma's brother. Then she and her family left Herat and no one knew where they were. Salma's brother had four children of his own and gave this child to Salma.

Adoption is emotionally difficult and legally complicated in Afghanistan. If the boy's family were to change their mind and come looking for him, Salma would lose her son the same way the woman in the wheelchair lost her only daughter. Legally, Salma has no rights over the child. Socially, so far, she is respected as the mother of this boy because she has told everyone that he is the child of her older brother, and her own mother enjoys being his grandmother. Salma is happy because she sees herself as building a secure future. But she wants to raise her son with honesty. She has decided that her son will know the truth about his heritage, even though all her family members have told her not to reveal it to others.

New Challenges in the Workplace

As Salma's home life became more settled, work became more difficult. She faced new problems as she rose in the NGO management. She was promoted as the head of the administration section and finally as the deputy director general when the head of the agency was out of the country for a while. But she found the work difficult. She needed to improve her English skills in order to write reports for donors in English. Also, she needed to learn advanced accounting and finance skills, but the agency told her that they could not spare her to go abroad and learn either skill.

When the agency was reorganized and the head of the agency left, instead of being promoted to the key position, Salma was moved out of management into a peripheral position. The reason given was that she did not have the necessary skills in English or finance. Her problem was that, in spite of invitations to travel abroad and learn these skills, her agency had not allowed her to take the time off. Then when it was time for promotion, they used her lack of skills against her. Although she had been running the agency in the absence of the founder and the executive director, there was a feeling in the agency that a woman should not be appointed to that post. Salma was demoted, and a man with no more experience than Salma was given the job.

Salma was depressed and angry, for she had worked hard to develop her skills and prove her worth in management and administration. She had recognized her weaknesses and tried to find ways to overcome them. But in the end cultural factors worked against her. Still she did not want to destroy the agency or harm it. Rather, she wanted to move forward by bringing her many management skills to her new posting and continuing to work. She was appointed as head of the education sector for the agency. There she faced another roadblock; this time it was from a woman whom she had hired and promoted. This woman told the new director that she would not work with Salma and submitted her resignation. Salma offered to step down rather than have this capable woman quit.

Neither resignation was accepted, and Salma was moved to a new department as the head and chief editor of a nationally distributed monthly Afghan magazine. Here Salma was her own boss again and could start shaping the magazine in ways she thought would be valuable. She changed the publication schedule of the magazine from monthly to weekly and added sections for women on health, mental health, coping with stress, cooking, fashion, and women's rights. She also added articles in the children's section, like information on sports, which she thought might interest adolescents. The magazine now has articles on politics, health, gender issues, and youth and children. Salma is looking at ways to use the magazine to promote civil society and democracy. At the same time, she ensures that her staff is gender balanced, aware of the needs of any disabled staff, and open to all ethnic groups in Afghanistan.

Thus, Salma did not give up when she was demoted. She was disappointed that she was not recognized for her efforts, that her work for this agency did not change the fact that she was a woman, and because of that she would not be given the position of director. Yet, in her new position her outreach and the effects of her work for the rights of women and minorities, for ethnic reconciliation and cooperation, and for civil society and democracy are greater than she could have imagined when she was "demoted."

Winter 2004

Salma never seems to be tired of her work. She is a little older, and I think much wiser, than she was when she first returned to Kabul. Her sense of herself is firm and grounded in the current political and social realities of present day Kabul. Her sense of her place in Afghan society, as an educated, working Muslim woman, is very clear to her and to anyone who meets her.

While people in the West think of Afghan women as victims, as isolated and disenfranchised members of their society, Afghan women, like Salma, have a different view. These are the women of the upper and middle classes, for whom education is the norm and political and economic activity is expected. It is ironic that when feminist activists in the United States were rallying around women whose lives were confined by the burqa—the cover that envelops and hides all of a woman and allows her to look out at life through an embroidered mesh face mask—women like Salma were donning the burqa in order to meet with and help poor women throughout their country.

Salma is one of the many bright, young, energetic women of Afghanistan who are working to move their country forward and to acknowledge the rights of women. But Salma is able to do this only because her family and community support her actions. Her ability to enlist their support in the context of rapid social and cultural change is impressive. Salma exemplifies the modern Muslim woman of a country in conflict and searching for peace. Her efforts to positively affect gender rights and women's needs are successful because she maintains a position that can be recognized in this Islamic society.

As I sit writing this in Kabul in 2004, fighting still continues elsewhere in the country. The United States is leading the fighting with much of the same rhetoric that was used by the Russians when they invaded Afghanistan in 1979—as many Afghans have pointed out to me. At the same time, Afghans say: "If we don't support this regime, the country will be torn apart by factional fighting." Nothing seems to change in Afghanistan. And it may never change if women don't have a greater say in how their society is run. There is hope for such change because of women like Salma.

CHAPTER 9

Growing Up Muslim in America: Dr. Anisa Cook *(African American Pediatrician)*

Frances Trix

Anisa, dressed in her dark coat, long skirt, and soft white scarf, or *hijab*, that framed her face, came in from the Children's Hospital later than she had planned. The *hijab* was pinned under her chin and on one side, and it fell gracefully across her shoulders and down her back. The dinner I had made for us would have to be warmed up, but it was so good to see her. I had seen her only once since she had come back as an attending physician at the Children's Hospital in a large midwestern city. Her work was demanding, with long hours and much responsibility. "I like the challenge," she said. And then she noted how fine the nurses at Children's were, and that the director was most supportive. Difficult cases were brought to Children's Hospital from a thirty-mile radius, and so her challenges were ongoing. As she spoke, I thought it was fortunate for the children and parents that she had come back. She had done her residency there and had not planned to return. But when she did return, she said there was a world of a difference being a resident and being an attending physician.

As we had dinner in my home, I asked Anisa what it was like growing up Muslim in America. I had known her for many years, but somehow this topic had never come up before. She received three calls from the hospital during the evening, one was for a case of a young person whose condition had changed rapidly. I marveled at her concentration as she instructed the resident who was in charge in her absence.

I had met Anisa many years ago when she was a teenager. We lived in the same apartment complex on the edge of a university town in Wisconsin. When I saw the good care she took of her two younger brothers, I asked if she would ever be interested in babysitting my young son who was between her brothers in age. She was interested, and my son liked her so much that he would ask me if I needed to go some place so Anisa could come over. They would have pizza together without cheese and pepperoni, because my son was allergic to dairy products and Anisa didn't eat pork. I found that I too liked talking with her, and I was thankful to her advice for sending my son later to the same high school she attended.

Anisa was the only Muslim in her high school, or so I thought. I had always seen her with her white scarf on; she always wore it tied under her chin, and long down her back. She was on the field hockey team. Actually she was captain of the team, and she played volleyball too. "Was it ever a problem with sports?" I asked nodding at her scarf. "Only once," she answered, "when a referee at a field hockey meet in Virginia had us all line up before the game. He wanted to make sure there were no earrings or barrettes that could hurt us. When he came to me he said, 'Let me see your hair.' 'You can't. It's religion,' I said, and he kept on moving down the line." In volleyball meets, some had suggested that it was important to stick to the uniforms that left players' arms bare. Anisa, of course, covered her arms, and it was left at that.

It turned out that she was not the only Muslim in her high school. There were two others: a younger boy who was quiet and hung out with the computer crowd, and an older girl. But the older girl did not wear a headscarf, and Anisa did not know her name or realize she was a Muslim until they met at a Muslim Student Association meeting her first year at the university. Anisa said she had thought she was the only Muslim in her high school because she was the only visible one. I asked if this had been difficult.

Anisa felt that she was accepted for what she was. But she added that she had started at the school before she began to wear the headscarf. When she had first come in seventh grade, her parents had spoken with the school officials so that she could perform her prayers in a private place. The school authorities let her use the sick room for prayers when there was no one in there. If someone tried to enter when she was praying there, the staff would say that Anisa was in there praying for a few minutes. But when she started wearing the headscarf a few weeks into eighth grade, being a Muslim suddenly entailed a more visual separation.

Anisa knew from her parents that she would have to wear the *hijab* when she reached puberty. Her mother, who is a nurse, always wore a headscarf. But Anisa had no sisters, and her Muslim girlfriends from earlier years had moved away. So when she had to put on the headscarf for the first time, she was actually frightened. The night before she went to school wearing the scarf, she called her locker mate to tell her about it. When she got to school that first day wearing the *hijab*, she went straight into the bathroom and cried. But she came out when people told her not to worry about it. And the people were supportive of her.

Later on, her mother told her about what she had said to her father, "She has her hair covered. Everything else will come." So her parents allowed her to play on sports teams as long as she was modestly covered. Swimming would have been another matter, but since there was no pool at the school, that did not come up. Her mother understood well how the scarf was a symbolic barrier. In her senior year in college, Anisa said that she had actually called her father and said, "Dad, I know you didn't force me, but I just want to thank you for telling me to wear my scarf during junior high and high school." She felt that it had protected her, even when she did not know it. And this, combined with not dating and not drinking, had saved her from much grief that she heard about from other girls.

Anisa's Family

I asked Anisa if people at the hospital realized she was an American Muslim or if they thought she was from overseas. Two-thirds of the five million or so Muslims in America are from overseas or are the families of people who recently immigrated from overseas, and one-third are African Americans. Anisa said that indeed people often asked her where she was from. She would respond, "Do you mean originally?" When they nodded, she would tell them with finality, "Philadelphia."

Both parents of Anisa are from Philadelphia, and she was born there and lived there until her family came to Wisconsin so her father could go to graduate school at the University of Wisconsin. I asked her if her parents had grown up Muslim the way she and her two brothers had. They had not, she said her mother had grown up as a Roman Catholic and had gone to a Catholic high school. Her father had also grown up as a Christian. But like a considerable number of African Americans of their generation, they had both become Muslim in their early twenties before they met each other. But unlike many

African Americans, Anisa's parents had gone straight to Sunni Islam.

Anisa explained that a more common pattern for African Americans was for them to come to Islam through "the Nation." What she referred to here was the movement that started in 1930 in Detroit, which then spread to Chicago, New York, and Boston, and became known as "The Nation of Islam." It was an early black separatist movement that emphasized hard work, frugality, discipline, and a conservative lifestyle. But it was more Islamic in symbol than in beliefs. Malcolm X (1925–1965) was a famous leader of the Nation until he converted to Sunni Islam. In 1975, the new leader of the Nation steered it toward Sunni Islam, although some members remained with the earlier black nationalist beliefs.

Some African American Muslims also make the point that between 10 and 20 percent of the people who were brought as slaves to North America and the Caribbean from West Africa were Muslim. Under the pressures of slavery, they were forced to become Christian. Thus, when the African Americans become Muslim, they are not converting but "reverting" to their former faith.

I asked Anisa if other members of her parents' families had become Muslim. A few had, including her father's mother just before her death, but on the whole, Anisa and her brothers remained "the Muslim cousins." They would go back to Philadelphia each summer and keep up their family connections, but her parents were the only ones in their families to both embrace Islam and raise their children as Muslims.

When I asked about Muslim influences on her parents, Anisa explained that they had never spoken of a single individual who had guided them to Islam. The 1970s was a period when many urban African Americans studied Islam. But at Temple University in Philadelphia, she noted that Professor Faruqi had exerted a wonderful influence on her father.

Ismail Faruqi was a professor of Islamic Studies and History of Religion at Temple University from 1964 to his untimely death in 1986. He was born in Jaffa, Palestine, during the British Mandate period and was educated at a French Catholic college and the American University of Beirut. In 1948, he immigrated to the United States where he completed his master's degrees at Harvard and Indiana Universities and obtained a PhD at Indiana University. Then he studied for four more years at al-Azhar in Cairo, which is the oldest Muslim university in the world. Professor Faruqi became a renowned scholar of Islam, as did his wife. He was also an early participant in ecumenical dialogues with scholars of other faiths. But Professor Faruqi also understood the importance for modern Muslims to know about their faith and the great Islamic civilizations of the past, and so he actively worked to establish schools of Islamic studies in universities in America and abroad. In addition, he was the founder or an early leader of groups like the Muslim Student Association and the Association of Muslim Social Scientists.

In Philadelphia, besides his regular courses, Professor Faruqi held Friday Circles that Anisa's father attended. When she was still a baby, her mother would dress her up, and then her father would take her with him to Friday Circles where she would be passed around and doted on, for children are considered as gifts of God. I think Professor Faruqi would have been most pleased to know that the young baby who had been at his Friday Circles had become both a pediatrician and a devout Muslim.

When Anisa's parents moved to Wisconsin so her father could go to graduate school, they became members of the local mosque. But since it was a university town, its members were largely international students from all over the world. Anisa remembers that she had a good friend who was Libyan but who moved away in sixth grade. By the time Anisa entered junior high school, there were only a few Muslims her age left in town, because they all had moved on after their parents had graduated from the university.

I asked Anisa if her brothers' experiences as Muslims in high school had been similar to hers. She thought that their experiences were different since, unlike her, they had no outward symbols of being a Muslim. "Only their names are Muslim, but many African Americans have unusual names. They had more to contend with people's prejudice against black males." The youngest brother was into soccer. The other brother was less interested in soccer, but after Anisa was in college when the family moved to Texas for three years, this brother also wanted to join the football team; it was partly because he stood out for his intelligence, and joining the football team would have eased his situation. It was more acceptable for a black male to be in sports.

Muslim Groups at the University and Medical School

When Anisa went to the University of Wisconsin in the early 1990s, she was one of the few Muslim women who wore the headscarf. This meant that when someone saw a white headscarf on campus, they would assume that it was Anisa. But things changed during her undergraduate years, and it had become the norm, by the year after she graduated, for Muslim women to wear the headscarf. This was influenced by events outside the university and by the Muslim Student Association.

But for Anisa in her first year at this large, somewhat impersonal university, what mattered most was a small group of Muslims, mostly of Pakistani background, who had grown up in America and who took her into their fold. They were a few years older than she was, but they had been students of her father at the mosque and so looked out for her.

At this time the Muslim Student Association there was in flux, as was the case all around the country. The Muslim Student Association was founded in Urbana, Illinois, in 1963, by Muslim students from overseas who were studying in the United States. By the late 1970s, it had become the largest and most active Muslim organization in America. But by the early 1990s conditions had changed. The organization was mainly run by graduate students at that time, and they said that there needed to be two Muslim Student Organizations: one for women and one for men.

But the Pakistani American students insisted that there should only be one group, and that women should also be included in it. Due to this difference of opinion, this group, which included Anisa, broke off and formed what they called the Islamic Circle. There were about thirteen of them and they would meet once a week for an hour. For the first half, someone would recite a passage from the Qur'an and interpret it in a practice known as *tafsir*. During the second half, they would talk about some issue related to Islam. They were mostly Sunni Muslims, but there was also one man who was a Shi'a Muslim. This experience of learning from each other was empowering, and the number of people who came to the Islamic Circle grew.

By the middle of the following year, the earlier leaders of the Muslim Student Association had graduated. And new Muslim students who had been members of the Muslim Youth of North America had arrived on campus; they were accustomed to Muslim organizations that included both women and men. This, coupled with the growing size of the Islamic Circle, led them to merge and again become the Muslim Student Association. The next few years were active years for the Muslim Student Association there, and they regularly drew thirty to sixty people to their weekly meetings.

When Anisa reflected back on these times, she noted that the people she met with each week in the Islamic Circle and in the Muslim Student Association became her first group of Muslim friends. When she was much younger, she had known a few Muslim girls of her age, but she was never part of a group of Muslim friends until this time. Besides being friends, being with them was also a learning experience; it brought her into regular contact with Muslims who had not grown up the way she had. She explained that many African Americans separated themselves from American culture when they became Muslim. Thus they maintained fairly strict separation between the sexes in public gatherings. However, such separation was not so strictly followed by the Pakistani and Indian Muslims. This had surprised her at first, but then she recognized that they were also Muslims, they too prayed five times a day, and such differences could be tolerated.

I asked her whether they dated. I was told that they did not date. But they all worked together on projects. And they were from a variety of ethnic backgrounds as well as a variety of Islamic backgrounds, from both Sunni and Shi'ite. Despite the differences, they were all there to make people aware of Islam and to understand what it was to be a Muslim. They also got together for holidays and for prayers. Despite their differences, they learned that they could all get along. And since they all lived on campus, they became like members of a family to each other. After the weekly meeting, the girls and guys would separately get together for pizza or coffee. Some were from wealthy families, and others were from working-class families. But at the university, they all became one big family.

Anisa contrasted this experience with what she encountered during her first two years at medical school. Again, there were Muslims from different ethnic and Islamic backgrounds, and they worked to meet regularly. The difference in medical school was that there were many who had no previous experience in dealing with Muslims outside their own backgrounds. In particular, some of the Muslim men in medical school, for various reasons, were not cohesive. But the Muslim women were.

Anisa became part of a core group of Muslim women who formed such a tight bond that they have maintained it to this day, despite marriages and children and other changes in life. They were all Sunni Muslims and all had grown up in the United States, but ethnically they were of Syrian, Pakistani, Palestinian, Indian, and African American backgrounds. Several had not previously been part of any Muslim group. One of them, who had been a homecoming queen in high school, started wearing the *hijab* partway through medical school, as did another. And another among them began to perform her prayers regularly.

When Anisa reflected on why these women were able to form such a tight group, she felt that it was because they had all decided to put

Islam first, whereas some of the Muslim men seemed to look on Islam more culturally. "Perhaps," she speculated, "we Muslim women came to recognize ourselves as Muslim first because visually, that is how society saw us." But it was the decision to wear the *hijab* that led to this in the first place. Anisa's core group of women looked on Islam as a religion that they practiced daily, not just took for granted.

A Year of Learning in Egypt

Midway through medical school, Anisa decided to take a year off to go to Egypt to study Arabic. In Cairo she lived first with an Egyptian family and then with a Sudanese family. For the first three months she experienced culture shock. As an African American, she had never seen herself as particularly patriotic, but after several months in Cairo, she came to realize just how American she was.

And to her surprise, she experienced more racism in Cairo than in the United States. As she put it, "Egypt and Sudan are right next to each other, and they say they are sisters, but they really are 'color struck' over there." Anisa said that when she spoke Arabic, they thought she was Sudanese or south Egyptian or just "this small African girl." I asked if the Egyptians had heard of African American Muslims. She had answered that they had not heard of them at all. And the racism against her stopped only when they found that she was American.

I too was surprised by this. Islam has a reputation for seeing all humans as equal before the overpowering oneness of God. And it is well known that the Prophet himself appointed a black man, Bilal the Ethiopian, as the first person to chant the call to prayer. When I asked Anisa about this, she agreed that Islam was not supposed to have any racial prejudice, and that it was only important that people were Muslim. But she also noted that people tend to take their baggage of prejudiced views with them wherever they go. She noted sadly that sometimes this plays out in mosques and the new immigrants may pick up the prejudices of the majority, or they may bring these prejudices with them from their home societies. "Unless they see themselves as Muslim first—but in any country these are the special people."

I suggested that it must have been a hard year. Anisa agreed, but she felt she had learned a lot, especially about herself. She realized she was at home as an American Muslim. And she also realized that home was where you should be willing to put up with nonsense, because there was nonsense everywhere in the world. She also learned that whether one was a practicing Muslim was dictated not by country but by person. She saw many things being done by people who claimed they were Muslim, but their deeds were utterly unacceptable for Muslims. She had not been exposed to such aspects earlier. She learned that just because a country was considered to be a Muslim country did not mean that one would find more Islam there.

At the same time, Anisa was impressed with the strong sense of family among the Egyptians. She also found it remarkable that Cairenes were willing to mediate disputes that arose between all types of strangers. And she learned more than Arabic from the teacher who taught her the last seven months in Cairo. He was a remarkable teacher and had an amazing drive for learning that would stay with her.

Medical School, Residency, and Return as Attending Physician

When Anisa came back after a year in Cairo, she came back determined to learn anything and everything. Perhaps it was a newfound appreciation of home, or the loss of an idealized notion of a Muslim land, or perhaps it was the model of her remarkable Arabic teacher whose drive for learning had so influenced her. In any case, she became more aggressive in her learning. She would search out those who could answer her questions.

During the last two years in medical school, the clinical years, Anisa did most of her rotations in a large urban medical center whose physicians, residents, and even patients were heterogeneous. But she also did some rotations in private suburban hospitals where the atmosphere was different. I asked if she felt there was prejudice against Muslims there. She described some dispiriting situations in which young medical students were treated as if they didn't exist. One of her fellow medical students who was an Asian noted how racist they were. When Anisa heard this comment, she realized that she wasn't the only one who had difficulties there, but she did not ascribe this to her being Muslim. And she had felt very much as an outsider at another private suburban hospital, both as an African American and as a Muslim. Other than these situations, she did not think that she experienced difficulties because she was Muslim. Her clear attitude toward learning may have averted such attitudes, for she had a no-nonsense attitude. And she had long ago decided that being Muslim would not impinge on her dressing nicely and looking professional while still covering her hair.

When she had to find a way to get around something in different situations, she managed to do it. For example, when she had to go into the operating room, she merely wore the scrub robes, which the technicians wore, to cover her arms, and put the little hat over her hair. There was also a situation in which obstetrics nurses criticized her for wearing t-shirts under her scrub outfit to cover her arms. So she used long sleeved shirts made of scrub material or wore jackets the way some nurses did. Here, she also recognized that the nurses would always find some reason to criticize pediatric residents. It was not personal.

One area that she did not adapt to, however, was the social one in which there was much alcohol. As Anisa put it, "Physicians drink like fish." As a Muslim, she did not drink, and so she was not in on the latest social events. But this allowed her to focus even more on her education.

During her three years of residency, she was chosen as one of the two chief residents. This was a distinct honor, but it also required her to serve a fourth year of residency. As chief resident in a large children's hospital, she had to represent the residents and act as the liaison for the administration. She said she had not wanted to be a chief resident, but the program director, who was Asian American, was most supportive of her. She also recognized that she had become more confident and that discrimination was a fact of life. If she was meant to be there as chief resident, God willing, she would be there. She had learned this wisdom from her mother, and it held her in good stead.

But her chief residency year was a year of learning for her in a way she had not anticipated. The other chief resident that year was a very popular white male. The difference in the way he was treated and the amenities he was given stood out. She watched "the old boys network" in action as some of the older male physicians gravitated to him. But she also realized that they were relating to him in that way just because he was "like them" socially, and not because of his skill as a physician. She saw more clearly that this was a different group of people from those who valued people for their skills and commitment; the latter were those who themselves possessed these skills and values. During the year, she also watched the male chief resident come down a peg or two when he realized there were times when he too had to prove himself, instead of just walking into the room and declaring himself as the doctor.

After completing her year as chief resident, Anisa took up a position as an attending physician in a town in another state. The people in that town were mostly Roman Catholic. But what mattered to

Anisa was that they were good family people for whom religion mattered, and they did not exclude others from their lives. She took the position there partly because of how genuinely kind the people were and also because they clearly valued her for her skills. When her mother and brother visited her there, they too remarked on the character of the people and their cordiality.

She said that her tenure there was difficult not because she was Muslim but because she was alone, she had to go there all by herself, and also because it was her first year as an attending physician. There was only one mosque in that town, which meant that everyone had to get along. And again there was a Muslim family there that looked out for her.

After a few months, she realized she wanted more professional challenges and started to look for positions elsewhere. She never thought she would return to where she had done her residency at the large urban Children's Hospital, but things just fell into place. People were pleased to have her back and welcomed her.

Islam in Her Life and Post-9/11 in America

Throughout Anisa's account of her life, she frequently remarked how fortunate she had been. She did not think that she had experienced much discrimination as a Muslim woman. But the way she had put on blinders, in order to focus better on learning, may have contributed to her outlook, and also the obvious quality of her character that was apparent when she was a teenager.

Islam is clearly central to her life; she is grateful that she is a Muslim and also grateful that she grew up as a Muslim. She sees that both wearing the *hijab* and avoiding places where people drink alcohol have saved her from the unfortunate experiences that she learnt about from her non-Muslim friends. She never had any bad experiences at parties and her education was not affected by vagaries of boyfriends. As a young person and as an African American, she has also seen where emphasis on the right clothes, the right jeans, and the right music in American life has left people adrift. In contrast, Islam places signal value on the family and community.

Being a Muslim in America also exposed her to other cultures. She is grateful for this and for the opportunity to learn what is good in these cultures and possibly incorporate some of this in her own life.

According to Anisa, Islam lays down limits and boundaries, but not to the point that she feels constrained. And she sees in Islam a corrective influence for the loss of values, like respect for the elderly, that has taken place in American life. Much that she learned from her parents was reinforced by being Muslim. Her parents would say, "If you want to be a good Muslim, this is how you should act." She learned to follow these teachings and to study them. Overall, Anisa remarked that she was thankful for these values as she grew older, and also thankful to those people who cherished these values.

I asked her if she thought that a Muslim girl growing up in America and who was not an African American would have had a much different experience from that of her own. Anisa did not think there would have been any difference as long as she did not have an accent and provided it was before 9/11. She feels that Arabs today have to contend with additional discrimination. But she remembered some experiences that her Arab friends had even before 9/11. For example, when she was eleven, she went to the movies with her Libyan friend to watch "Back to the Future." And in the movie, the reason for the person going back to the future is due to the presence of Libyan terrorists. She also remembers how assimilated many Arab American young people used to be until they went to college. And there were difficulties faced by Muslim families arriving from overseas when they experienced the contradictions between Islam as they had practiced it back home and life in America.

In mosques in America, Anisa noted that there was often an immigrant-indigenous split. But the events of 9/11 made these communities realize that they needed to have more of an American presence in the mosques. It is accepted that Muslims who came to America from overseas brought with them knowledge of Islam as they lived it. But that fact alone did not guarantee that they were the best Muslims, or the best representatives of Islam for the community. "The experience of 9/11 made Muslims in America realize that we need to be with the society here, we cannot just be in our own little world. We need to reach out to others in the society." Anisa, her parents, and brothers all epitomize this deeply held yet outward looking Islamic faith.

CHAPTER 10

A Javanese Muslim Life of Learning: Professor Dr. Haji Mohamad Koesnoe (*Indonesian Legal Scholar*)

R. Michael Feener

It was a long, blindingly bright ride out of Surabaya and past the salt flats—mile after mile of barren, bleached plots where the sun had seared away the seawater. Eventually, however, we found ourselves climbing up from the coast into the cool green of the hills. We were on our way to Gresik, the resting place of Sunan Giri, one of the *Wali Songo* (Nine Saints) of Java who are traditionally credited with bringing Islam to that Indonesian island during the fifteenth century. I was invited to visit the gravesite of this Muslim saint by my host and teacher Professor Dr. Haji Mohammed Koesnoe. Although he carried this distinguished set of titles before his name, he preferred that I simply refer to him as "Pak," an Indonesian term for father that expresses both respect and affection. Having suffered a severe stroke just a few years earlier, Pak Koesnoe was unable to climb the steep stone stairs leading up to the shrine and cemetery. Instead, he sat in the shade below chatting with the sellers of prayer beads and Pepsi who daily set up shop just below the main gate. He told me to go up and to take my time.

Passing through the elaborately carved stone gateway, I began climbing the steep flight of well-worn stone stairs leading up the mountainside, passing dozens of graves that rested under a lush canopy of trees on both sides of the steps. At the top of the hill was a mosque built in the classical Javanese Muslim fashion, topped with a stacked pyramid roof of red tiles rather than a dome in the Middle

Figure 4 Mohammad Koesnoe with Author

Eastern style. Nearby was a smaller structure, the wooden walls of which were carved with intricately stylized floral patterns. I crouched and crawled through its small wooden door, which was flanked by a pair of *nagas*, serpent-like creatures that have guarded sacred places in Java since before the arrival of Islam to the island. Inside I found myself in a crawl space that ran around a partitioned central chamber housing the grave of Sunan Giri. There were half a dozen others already inside, most bowed in silent meditation or prayer, reciting pious formulas in Arabic and repeating litanies of the ninety-nine most beautiful names of God—a practice popular among Muslims in many parts of the world. As I entered, one of them was placing an offering of tropical flowers on a tray kept near the gate of the tomb. I sat for a while in one of the corners, taking in the cool, dark, sweet-smelling peace of the place. After coming back outside, I passed by families picnicking on straw mats and in small pavilions scattered across the upper cemetery, with parents and grandparents snacking and sipping tea as children laughed and played nearby. As I descended the last few steps and met up again with Pak Koesnoe, he turned and asked me what I thought of the place, I responded, "Beautiful." He said, "That is also Islam."

Pak Koesnoe's vision of Islam has had a tremendous impact on my thinking about religion and society, and in many ways it was because of him that I had made this first trip to Indonesia. I first met him when I was an undergraduate at the University of Colorado, Boulder. He had come there as a visiting Fulbright professor in 1990 and co-taught a course with Professor Fred Denny on Islam in Indonesia. Before that time, I had very little exposure to Islam and Muslims, and the course was an eye-opener to me. There at the University, I first came to address some of the major questions that have occupied my mind ever since: questions about the relationship between religious ideals and the practical realities of the human cultures in which they are situated and lived.

Pak Koesnoe's lectures engaged with issues of global scope from a distinctly Javanese perspective. His teachings challenged both Muslim students and Western academics to reevaluate given models of understanding Islam through his method of discursive explorations of what it might be like to view Islam from a Southeast Asian perspective, rather than via a perspective situated from either the Middle East, or the Midwest. This orientation toward the appreciation of Islam in local contexts had developed out of a life of careful study and reflection, during the course of which he had traveled around the world. It started, however, from a strong sense of the place where he was from—the island of Java, in the Indonesian Archipelago.

Student Years to Indonesian Independence

Pak Koesnoe was born in the East Javanese town of Madiun in 1928, but his perspective on the world was by no means provincial. As a teenager he began traveling throughout Java and the surrounding islands of the Indonesian Archipelago for study; this was the first step on a journey of learning that he pursued for the rest of his life. For most of his childhood, the Indonesian Archipelago had been subject to Dutch rule, and they ruled the islands of the East Indies as colonies of the Netherlands. During his student years, much was happening in the Netherlands Indies. Throughout the 1930s and 1940s, this society was in the midst of a long period of modernization. Some of the most important developments related to this social transformation occurred in the field of education, with new schools employing new kinds of curricular materials made possible because of recent local revolutions in print technology. The first printing presses in the country began to make a broader range of reading matter more widely available, allowing a growing modern reading public to read works from Europe, the Middle East, South Asia, and throughout Indonesia itself. At the same time, this growth in new ideas was accompanied by the development of new institutions through which the peoples of the archipelago came to organize themselves in modern ways to advance their new ideas, religious understandings, cultural interests, and political aspirations.

With the outbreak of World War II, the Japanese military occupied most of the Indonesian archipelago, killing or imprisoning much of the ruling Dutch population in the process. In their place, the Japanese command mobilized local leaders to support them in their administration of the islands. During this period Pak Koesnoe attended Japanese wartime schools and even won a student prize in a Japanese-language speech contest. However, with the collapse of the Japanese administration of their "Great East Asian Co-Prosperity Sphere," the local leadership that had been mobilized under the Japanese started to become the basis for more organized movements for Indonesian independence. Following the defeat of the Japanese at the end of World War II, the Dutch attempted to take back their colonial possessions in the archipelago. They were eventually repulsed by an armed Indonesian resistance made up of volunteers from nearly every sector of society. Pak Koesnoe joined in this armed struggle against Dutch reoccupation, and received decorations for his service in the campaign for Indonesian independence.

Searching for a Legal System for the Diverse State of Indonesia

After the war, Pak Koesnoe chose to serve his country by devoting his energies to searching for a legal system to ensure justice for Indonesian citizens in his newly independent country. During the early years of the Indonesian Republic, its citizens actively debated fundamental issues relating to the way in which the identity of this new nation was to be defined. During that period, conflicts between nationalists, socialists, and Islamists were most pronounced in debates over the constitutional basis for the state and the model of law to be promulgated by it. A certain degree of compromise was achieved under Indonesia's first president, Soekarno, with the promulgation of *Pancasila* as the official ideology of the Indonesian state. *Pancasila* set a program for the creation of a society based upon the values of belief in one god, human dignity, social justice, national unity, and "guided democracy." Along with these five fundamental principles came the new national motto of *Bhinneka Tunggal Ika*—roughly translated as "unity in diversity." This aspect of *Pancasila* attempted to provide a framework for the vast internal pluralism of Indonesia, as an archipelagic nation of thousands of islands and home to peoples of hundreds of different languages and cultural traditions.

Pak Koesnoe completed his first degree in law at Jakarta in 1955. His studies concentrated on the field of customary legal cultures in Indonesia, referred to as *adat* law. The objects of study in this field were the traditional legal communities of Indonesia's diverse ethnic groups. However, in the modern period these "traditional" legal systems were studied at universities in rather modern ways, and it was part of an academic tradition in the Netherlands to approach *adat* law by means of social-scientific inquiry. In these studies the focus was on the collection and classification of local legal rulings and details of institutions to support the administration of colonial courts in the Netherlands Indies. After gaining independence, however, many Indonesians began to look toward ways in which the traditions of *adat* might inform new models of law for modern Indonesian society.

In 1960, the Indonesian National Assembly passed a resolution that emphasized the importance of *adat* as the foundation for the legal system of the new nation. At about the same time, Koesnoe began to undertake further studies of *adat* law and the intellectual world in which Dutch discourses on it had been formulated. Through a critical rereading of the scholarship of the colonial era, he came to realize the need for a thorough reevaluation of *adat* law and its academic study in postindependence Indonesia. At the center of his new studies on *adat* law was a fundamental shift of disciplinary perspective. Pak Koesnoe argued that the primary approach to *adat* law should not be through the social sciences, as was characteristic of much of the Western scholarship, but rather by means of legal studies and comparative jurisprudence. Beyond just theorizing on such a position, he also conducted field studies of Indonesian legal cultures along these lines, both in his native Java, and among other ethnic groups of the archipelago, including among the Sasak on the island of Lombok.

"Flying Professor:" Across Indonesia and the World

Over the years, Pak Koesnoe continually refined his thinking on issues related to *adat* law, not only in the seclusion of his study, but also in conversation with other Indonesian scholars and students across the vast archipelago. From 1956 to 1986, he served as one of the "Flying Professors" of his new nation who taught on several campuses simultaneously in an attempt to fill the great educational needs of the country despite its small number of trained teachers. This was a period of intense activity in teaching and scholarship for Pak Koesnoe, during which he constantly crisscrossed the country. In the process he not only shared his learning and insights with others, but also gained deeper understandings of the complex interrelations of various local cultural systems in different parts of Indonesia. During this phase of his life he taught at several institutions in Java, Sumatra, and Sulawesi, which ranged from secular faculties of law to the newly established "National Islamic Studies Institutes," referred to in Indonesia as the IAIN. In his postindependence home base of Surabaya, East Java, Pak Koesnoe was a founding professor of the Education Faculty at the IAIN Sunan Ampel when it was established in 1961. Since the 1960s, the expanding number of IAIN campuses in cities across Indonesia have become important centers for developments in how Indonesian Muslims have come to understand Islam in relation to their society.

While teaching, Pak Koesnoe was constantly studying as well, and in 1965 he earned his PhD from Airlangga University at Surabaya. Upon completion of his degree, he moved on to a new period of intellectual development. Soon thereafter, he also embarked upon his first adventures of international travel, thus beginning a stimulating and productive international scholarly career that spanned three decades. His first trip abroad was to the Netherlands in 1968–1969, where he was a visiting professor at the University of Nijmegen. Thence he moved on to a similar post at the University of Hull, England. Over the decades that followed, Pak Koesnoe expanded his itineraries to include visiting professorships in Saudi Arabia, Australia, France, the Philippines, and the United States, where he served as a visiting Fulbright professor. In addition to these travels, he was often made shorter trips to give lectures and attend conferences around the globe, including visits by invitation to Egypt, India, Pakistan, Thailand, Japan, Singapore, Malaysia, Italy, Germany, Belgium, and Switzerland.

In the course of his travels, Koesnoe continued to debate and refine his thoughts by incorporating elements of various perspectives into his own critical reflections on *adat* law, and at the same time he inspired others around the world with his perceptions and incisive analyses of issues related to law and society. Koesnoe also produced dozens of publications in Dutch, English, and Indonesian. These writings dealt with a wide range of topics, from specialized reports on aspects of *adat* law in various Indonesian cultures, to discussions on the Indonesian Supreme Court and the concept of authority. He also explored the relationships of Islam and Muslim religious ideals to various institutions of Indonesian culture through papers that covered subjects such as Islamic charitable land endowments and inheritance law.

However, my introduction to Pak Koesnoe's thoughts on Islam and *adat* came not through his many publications, but rather through more direct and personal contacts in and out of the classroom at the University of Colorado. During the semester in which I first got to know him in Boulder, Pak Koesnoe, his wife, and his research assistant Ibu Gien hosted regular get-togethers with the Indonesian community in Boulder. Because of my keen interest in his class, I was invited to join these gatherings, and there I made many of my first Indonesian Muslim friends. At the end of the year, I was invited by some of them to visit Indonesia in the summer, and during my visit there I stayed for several weeks at Pak Koesnoe's home in Surabaya. It was during this first visit to Java that I made the trip to the tomb of Sunan Giri. That was just one of the many places where I learned from Pak Koesnoe about Islam and the way in which it was lived by Muslims in Indonesia.

Toward a Complementary Relationship Between Customary Law and Islam

Many twentieth-century discussions on Islam in Indonesia were framed in terms of a conflict between "Islam" and "local practice" (*adat*). A number of prominent Western scholars used this dichotomy to demonstrate that Islam was not deeply held by most Indonesians—that in a sense Indonesians could thus be viewed as being "not as Muslim" as their coreligionists in the Middle East. Such observations have been used by adherents of various political positions to argue in favor of the unique nature of Indonesian Islam. Complementing such views from the outside, a number of Indonesian Islamic reformers have themselves used analogous understandings of Indonesian Muslim societies to argue for the "purification" of Islamic ritual practice from what they saw as the "contamination" of local pre-Islamic cultural practices. However, such movements for religious reform, and the cultural situations to which they responded, are in no way unique to Indonesia.

In fact, in many parts of the Muslim world, including the Arab lands of the Middle East, much of local customary law had come to traditionally be considered "Islamic" as long as it did not support any actions that were clearly prohibited in the text of the Qur'an or the hadith of the Prophet Muhammad. As Islam expanded, in many places local tradition and conceptions of Islamic law grew so intertwined that a sharp distinction between them was not widely perceived by the people in such societies. The spread of Islam often involved the incorporation of local legal practices into the law of Islam, something that can be seen in the history of Muslim societies ranging from the early Islamization of Persia in the seventh and eighth centuries to the spread of the Ottoman Empire into the Balkans—the latter, coincidentally, occurred during roughly the same period as the first major expansion of Islam in the Indonesian Archipelago.

As in all Muslim societies, Islam in Indonesia is integrated into local practices of everyday life which, as mentioned above, are discussed in Indonesia in terms of *adat*—a word derived from the Arabic for custom. Much European and colonial scholarship and Islamic reformist rhetoric tended to conceptualize *adat* as oppositional to Islam, despite the fact that many Indonesian cultures themselves had traditional formulas that expressed a more complementary relationship between the two. As the Minangkabau Muslims of West Sumatra expressed it, "*Adat* is based upon the Law, and the Law is based upon God's book (the Qur'an)." Traditional Indonesian Muslim sayings like this one capture some of the dynamics by which the rather minimal requirements of Islamic law are elaborated in local cultural expressions that make such observances an integral and meaningful part of one's life as a Muslim.

Adat as Cultural Practice and Ethical Mode

On one level then, *adat* is used with reference to particular cultural practices. For example, in some parts of Indonesia it is considered a traditional practice by Muslims to eat a steamed rice dish called *ketupat* at the end of the fasting month of Ramadan. The formal requirements of Islamic law stipulate that Muslims should fast during Ramadan and that the completion of these observances at the end of that lunar month be marked with prayer. However, the details of the celebrations that accompany this observance—including what foods are to be eaten—are left to Muslims themselves to determine, and thus they can vary considerably across cultures in different parts of the Muslim world. Through his rich range of experience in, and thoughtful reflections upon, many of the diverse cultures of the Indonesian archipelago, Pak Koesnoe developed a deep appreciation of local cultural forms of Islamic practice and Muslim social life. This was clearly observable to anyone who spent time with him, as seen for example in his fondness for practices such as visiting the tomb of Sunan Giri.

Beyond such particular cultural practices, *adat* can also be approached on a broader conceptual level. At the heart of Pak Koesnoe's vision of *adat* law lay a sense of ethics as a moral force for the organization of society. In this he placed great importance on the role of education and the teaching of culturally sanctioned values, in which Islam had come to play vital and complex roles. His formulation of *adat* law emphasized not only its traditional nature but also its supple and dynamic aspects that served to keep the tradition alive and meaningful for generations of people in changing times. He thought of *adat* law as consisting of those aspects of tradition that served to express cultural conceptions of justice with regard to social relations. According to Pak Koesnoe, the material of *adat* law and its study was thus largely of an unwritten nature, but in which religious beliefs function as an intrinsic part. Thus, he argued that an investigation of the role of *adat* law in the governance of a given community should focus on the ways in which the general principles of *adat* embedded within the culture are interpreted by authoritative persons in the community and administered within its context.

Pak Koesnoe presented his thoughtful formulations of Islam in terms of the relationship of Muslim religious tradition to the conditions of local lived realities. He had a deep regard for the place that Islam had come to play in the historical development of the Javanese worldview and recognized that many elements of Islamic law had integrated themselves deeply into local *adat*. His studies of Javanese conceptualizations of Islamic ideals of *musyawarah* (consultative decision making), for example, explored the ways in which elements of Islam had deeply integrated themselves into Javanese and Indonesian culture more broadly. Pak Koesnoe looked to such things with a view to understanding what they did to establish and maintain harmony and social justice, while also thinking about what role such ideas might continue to play in modern Indonesian society. Such positions contradicted much earlier scholarship on *adat* law. Despite considerable criticism from scholars and jurists more comfortable with the earlier established approaches, Pak Koesnoe persevered while continually refining and defending his positions. For he believed that a reformed and reinvigorated study of *adat* law could serve Indonesia in its quest to establish a just and culturally grounded legal system according to which the country could best administer itself after having achieved its independence from colonial rule.

The Legacy of Pak Koesnoe

Over the decades that followed, however, *adat* gradually lost much of its appeal as a formal source of law for many Indonesians. National debates over legislative issues increasingly neglected *adat* law and instead focused on Western-style constitutional issues or the formal implementation of Islamic law in ways that would freeze the ever-evolving debates of Muslim jurists on the *Shari'a* into a compilation of fixed statutes. Nevertheless, many Indonesian judges continued to consider *adat* values as a factor when making decisions on issues ranging from land use to inheritance. It is in relation to such developments that one can understand Pak Koesnoe's conscious decision to approach and discuss *adat* law, not from the direction of formal legal practice, but rather through education. His work in this area had a significant impact, as a number of his students went on to become judges working in various parts of the country.

Although he chose the life of a professor, rather than that of a practicing jurist or legislator, Pak Koesnoe's Islam was not an intellectualized abstraction. For him, Islam, Javanese *adat*, and a modern, international academic career combined in a life of careful, critical thought and regular practice. In all the times I was with him over the years, I never once knew him to miss one of the required five daily prayers. Once, while we were preparing to drive out to visit one of Surabaya's historic mosques and things got to the point when everything was ready and we were set to leave the house, Pak Koesnoe noted the time and decided that we should wait for another forty-five minutes or so before heading out. The reason for this, he noted, was that the midday prayer time was coming up within the next half hour or so, and he wanted to make sure that we would not miss it while in transit. From my studies of Islamic law, I knew that according to several Islamic scholarly interpretations, some of the five daily prayers might be combined in certain circumstances if one happened to miss the scheduled time. I suggested that since we were going to a mosque anyway, he could simply combine the noontime and mid-afternoon prayers when we arrived there. He replied that although the letter of the law did make such provisions, he preferred not to avail himself of them. When I asked why, he responded, "The appointed times of prayer are blessings from God. You might think of them like little creatures that descend at the scheduled hours and perch upon your heart. They stay only a short while, and if you neglect that chance to pray they are gone forever—blessings lost." For Pak Koesnoe, combining prayers might preserve the form of the ritual and the letter of the law, but at the same time it would cause a loss of their substance and spirit.

Thus even after the stroke that physically prevented him from performing the prostrations of *salat* (Muslim prayer), he excused himself every time the call to prayer was heard and, seated in a chair facing in the direction of Mecca, recited his prayers while bowing his head. Pak Koesnoe was also meticulous in his fasting, and he was known for being generous in his charity above and beyond the formal prescriptions of *zakat* (mandatory almsgiving). Pak Koesnoe lived modestly and with a great sense of charity and beneficence toward the community. In all the time that I spent as a guest in his home in Surabaya, there were only few days in which at least one visitor did not arrive at his door. People from nearly every segment of Indonesian society came to visit him, and there they would receive not only material hospitality, but also generous advice and personal references that could help them in various ways.

In the conduct of his personal life, Pak Koesnoe was an outstanding model of Muslim piety. His ideal model for a law suitable for the cultural contexts of modern Indonesia was a far cry from that of those who advocated formal implementation of various constructions of Islamic law (*Shari'a*). While Pak Koesnoe upheld the importance of the ethical values of a traditional society that incorporated much from the teachings of Islam, he did not stress the necessity of any specific institutionalization of formal Islamic practice. For he saw the greatness and power of Islam in the dignity and beauty that its ideals had brought to Javanese life, rather than in the symbolically charged institutions that many modern reformers had come to exalt as the essence of the tradition. Of course, such a position exposed him to criticism from those whose visions of Islam and its place in Indonesian society emphasized the importance of implementing certain formal elements of Islamic law.

Pak Koesnoe's own Islam, one might say, saw the substance of lived Islamic ideals as more important than any particular manifestation of their essentialized forms. Since the 1980s, debates over Islam between "formalists" and advocates of more "substantive" approaches to the integration of religion into modern life have been prominent public issues in Indonesia. In his emphasizing of education and ethics, rather than institutional legislation, and in his discussions of *adat*, Pak Koesnoe lent a quiet but compelling voice to those who appeal to values like humanity, independence, and social justice as the foundations in their formulations of what it means to be both Muslim and Indonesian. Since he placed such great emphasis on the importance of ethics and cultural values in his thinking about law, the success of Pak Koesnoe's life's work cannot be adequately appreciated in terms of immediately visible products, such as the production of a written code of law or the creation of particular legal institutions. Rather due to the very nature of his work, his influence was exerted in less tangible ways on the thought and behavior of the many people who came in contact with him during the course of his life and were inspired by him as teacher.

Pak Koesnoe and I would often discuss ethical and legal issues in the course of the daily afternoon walks that he took as part of his stroke-recovery physical therapy. Since the sidewalks of Indonesian cities are notoriously ill paved, he had taken to walking inside shopping centers, which were easier terrain for his recovering legs to traverse and with the additional benefit of being air-conditioned against the tropical heat. Inside these spaces we passed by an eclectic mix of Western chain stores, boutiques specializing in the latest modest fashions for pious Muslim women, and snack stands selling tropical fruit juices and fried soybean cakes served in paper wrappers with hot green chili peppers and plenty of salt. As we walked, we navigated through the crowds that comprised people from diverse segments of Indonesian society, young men wearing Muslim skull caps with their jeans and t-shirts, older women dressed in local batik cloth, and others wearing imports or imitations of styles current in Japan, Hong Kong, and the West. Through our conversations during these walks, I learned even more than I did reading in his large home library, which was stocked with hundreds of books written in English, Dutch, German, and Indonesian. The global scope of his learning was a testimony to the extent to which he comfortably moved constantly between his deep grounding in local Javanese tradition, modern European thought, and an impressive Islamic intellectual cosmopolitanism. However, his erudition was worn lightly, and on his death in 1998 the world lost not only an accomplished scholar but also, and more importantly, a great teacher.

CHAPTER 11

A Hometown Artist of the World: Sabah Naim *(Visual Artist in Egypt)*

Jessica Winegar

Sabah Naim is now one of the few visual artists from Egypt who has become internationally recognized. You can find her on the Internet: she has exhibited in the famous biennial exhibitions in Venice and Havana, and in many other venues throughout the Middle East and Europe. Her work has sold like hotcakes, and the critics love her. But when I first met her just a few years ago, she was almost completely marginalized in the local Cairo art scene. Her work had been ignored by some of the most important Egyptian curators at the time. They dismissed her as an "ignorant girl" who didn't know anything about the cosmopolitan world of art that they fancied themselves to be part of. Sabah might not have been "cosmopolitan" in their sense of the word. She lived in a poorer neighborhood of Cairo where few people had completed even secondary education, let alone knew much about the visual arts. She did not speak any Western language and had never traveled outside Egypt. Like many young women her age, she wore a headscarf and talked openly about religious belief and practice. To the Egyptian curators, members of an older generation of elites, this last fact represented all that had gone wrong in Egypt and in many other Muslim-majority countries. But what they did not realize was that Sabah was cosmopolitan in a different way. Her drawings on photographs, sculptural collages of newspapers, and video art could transcend cultural boundaries whereas other artists' work could not. Rather than being a symbol of cultural regression or backwardness,

Figure 5 Sabah Naim: Collage

Sabah Naim was actually part of the vanguard of a new kind of Muslim artist. She was a hometown girl who made it big with her own artistic vision, not someone else's, and she had done it without giving up or apologizing for her way of life or her religious practice.

Vanguard of a Generation

Sabah Naim was born in Cairo in 1967. This was an important year for Egyptians and for many Arabs because it was in this year that they lost a war against Israel that they thought they could win. The Arabs bemoaned not only the Israeli takeover of various territories belonging to several Arab states, but also the end of an era in which it was largely believed that the newly independent Arab countries would join together in common interest and avoid being controlled again by any foreign power. After 1967, a new era began in Egypt. Many intellectuals ceased their activities or left the country disillusioned. This created a space for Sabah's generation to emerge as an effective force in cultural life. Sabah and many of her peers are distinct from earlier generations of artists and intellectuals because they were among the first to benefit from the free higher education policy instituted by President Nasser after Egypt's independence from Great Britain. Sabah grew up knowing that her family's limited income would not prevent her from pursuing a college degree and, later, a Ph.D. She also knew that she would be the first member of her family to get an advanced degree. Sabah and her family were also among many who benefited from Nasser's attempt to destabilize the class hierarchy and to create a new middle-class workforce for the modern nation. What this means is that Sabah is both highly educated and rooted in her unpretentious (and religious) social background. When she has an art opening in Cairo, she is able to easily chat with curators and art professors and then show her family around the exhibition. I always loved going to Sabah's exhibitions because they drew the kinds of people one usually never sees in art galleries anywhere else—people like her mom and dad, her siblings, and her neighbors.

Sabah was in the vanguard of her generation not only for her unashamed mixing of disparate social worlds but also for her artistic approach. After 1967, the Egyptian government gradually opened the country to Western investment and consumer goods—including everything from Chevrolets to Sony electronics to McDonald's hamburgers. Sabah and her peers thus grew up in a material world radically different from that of the previous generation. They also began their careers in the mid-1990s—a time when Egyptians started to have greater access to advanced technology and the intensified global circulations of media and products. Sabah is one of the first artists whose work clearly reflects these changes in Egyptian society and its relationship to global markets and culture flows. She does not do traditional oil painting or bronze sculptures, but rather uses new digital technologies and works with Egyptian and foreign news media. The images she selects are not idealized visions of a premodern Egypt filled with palm trees, camels, and pyramids. Rather, they are of modern, industrialized, urban life.

The majority of older generation Egyptian artists, critics, and curators—both men and women, in one way or another—express their disdain at the fact that Sabah does not sever herself from her social background or her contemporary context. There are those for whom contemporary urban life is an aberration of the "authentic" Egypt. To them, real Egypt is ancient Egypt, the land of the Pharaohs and King Tut. They also find real Egypt in the countryside, and in what they assume to be peasants' timeless customs and traditions. The urban is either too "Westernized" or a degradation of the "authentic." Many of these individuals also criticize Sabah's use of new media and technologies as too Western and accuse her of trying to imitate American and European artists. Sabah counters her critics by saying that Egypt is not only found in the countryside and in

Pharaonic history, but it is also expressed through satellite television and the Internet, and by wearing blue jeans or drinking Coca-Cola. She once told me that as an Egyptian artist she could not build her future solely based on the past. She had to include the present as well. The "present" to her is the proliferation of media and the changes in urban life that she portrays in her art.

Some older generation artists favor her aesthetic perspective. But they want to deny her the other part of her "present": the religious practices and social mores common to her social background, which have shaped Sabah as she is today. They would have probably loved these art works if they were made by a woman who spoke in English or French, did not talk openly of worshipping God or praying, wore short sleeves and makeup, and coiffed her hair, and did not bring her working-class family to her art openings. These are the things that brand her as hopelessly ignorant, traditional, and tasteless to the other older generation artists. Little did they expect in the mid-1990s that these are precisely the factors that would make her work transcend cultural borders so successfully.

People in Their Worlds, in Our World

When was the last time you were out in a public place and stopped to really take notice of the people around you? The next time you are at the bus stop, the grocery store, or the park, or in any other public space, take a minute and try to see people in their worlds in the way Sabah Naim's art suggests we can. The vast majority of her works over the past several years are pictures of Egyptians in public spaces, taken on her digital camera or clipped out of the newspaper. They are captured, as Sabah says, in their own individual worlds but simultaneously out in the public world. Many of her subjects are by themselves—waiting for the bus, reading the paper, smoking a cigarette. There are often other people milling about around them. Some scenes are of a pair or a group of people standing together or walking through the streets of Cairo—each silent, in their own worlds. Once Sabah has selected an image, or taken a picture or video of her subjects, she then draws stars, dots, curlicues, spirals, or flowers over parts of the image. These are most frequently done in gold and silver. Sometimes she also covers up whole sections of the image with solid blocks of color or with black hatch marks. The effect of these additions is to highlight the emotional states and gestures of some of the subjects, to focus on some of the interactions in the frame and elaborate on them, and to call attention to the subjects' relationship to their environment.

Sabah's artworks show scenes that are immediately recognizable to anyone who has spent time in Cairo. They are not the Western stereotypes of Muslims or Middle Easterners. There are no bearded mullahs, or women only in black robes, or men with checkered scarves wrapped around their faces with guns raised in the air. Nor are they the images of the rich and powerful Arab leaders, businessmen, and entertainment stars that grace both Western newspapers and local fashion magazines. For the most part, Sabah selects images of ordinary Egyptians going about their daily lives, doing normal activities such as waiting for the bus, walking to work, coming home from school, buying things from vendors, chatting with a friend on a street corner. They are recognizable to Egyptians because these are familiar things that people do in familiar environments and wearing familiar clothes. The images show worn-out benches on the sidewalks, peeling advertisements that were once pasted on corrugated metal store fronts, parked Fiat cars, small rubber trees, signs for the subway station, and big city buses. Egyptians will recognize the middle-aged government employees in safari suits, older women in their overcoats and colorful headscarves, men in the long robes called *galabiyyas*, young men in pressed dress shirts and slacks, and schoolgirls in their uniforms and backpacks. These are more typical of the scenes found in cities of the Middle East than the images one sees on the nightly news in the United States. Sabah chooses these images from her own environment. The people in her images could be one of her brothers en route to his job at the bus company, her sister going to high school, her mother on her way to the market or to see a doctor, or her father going to do business with other merchants like him. They could also be her coworkers at the College of Art Education, or her fellow artists running errands around town. These images are of the same streets she walks, the same buses and subways that she takes to get from her family's home to the college, then to the gallery that shows her work, from there to the paint store, to the coffee shop to visit with friends, and then to the mall where she loves to buy clothes.

At first glance, Sabah's world may seem very different from yours, given that she lives in an Arab city of eighteen million people, most of whom live on less than five dollars a day. But I believe that her intention to express both the specificity and universality of the human experience comes through in her works. They enable any viewer to see these people as individuals living valuable lives, rather than as an anonymous mass of fanatics who, as the pundits like to say, "don't value life." The artworks also enable the viewer to connect with these people and to recognize a bit of themselves in the images, rather than to see the people in the images as totally foreign—or worse, as enemies.

The process by which Sabah creates these works shapes their reception. She begins with empathy. When I asked her how she selects her images from newspapers, she said,

> I choose according to my own feelings. I find feelings within me expressed in the papers. [For example,] I chose one of the photographs because I felt the feelings of the man portrayed, and I felt that I could add things to it to capture his state.

The same goes for the photographs and videos that she takes. The empathetic connection with her subjects challenges the dominant Western modes of making images of the Middle East, which depend on establishing a distance between the Western photographer or viewer and the Middle Eastern subject. The distance created by the Western "gaze" of the "Other" enables Middle Easterners, and Muslims generally, to be imagined as exotic, fanatical, terroristic, and so forth. Such imagery was an important part of European colonization of the Middle East and still plays a central role in Western military interventions in the region. Sabah's selection of images challenges this dominant visual apparatus and all that it entails.

Sabah then performs her manipulation on the work she selects. The use of panels of color and the addition of stars, circles, flowers, and dots help create an atmosphere of intimacy and empathy because of the beauty and unthreatening nature of these symbols, which are pleasantly rendered in gold and silver. By covering up parts of the picture with symbols and blocks of color, Sabah also draws the viewer's attention to certain people, expressions, and interactions, rather than seeing them as a threatening mass. By using such symbols, which have been found in ancient and contemporary artistic traditions from all around the world, Sabah also draws on their universality. They are symbols that we all share. The circle, one of the most common symbols in these photographic works, is also found in her sculptural assemblages: she takes pieces of newspaper, coils them or rolls them into a ball, and then glues them side by side on a board. Sabah talks about the circles as being those of the individual self. But she is also interested in what happens when they are made repetitive—either through rendering many dots and circles on photographs or through coils and balls of newspaper. Sabah spoke about her inspiration for this repetition:

> I thought about becoming more regular, a person among others. And more repetitiveness, and that I am only one like the others. We are all people...from afar we are the same, but when you come closer we are different in the way we look, our way of thinking, and our way of dealing with other human beings....We are circles inside of each other, but inside each circle there is a world that is so distinctive from the one next to it.

I think it is partly this tacking back and forth between specificity and universality, between the individual and the shared, that makes Sabah's work translate across cultural boundaries. Viewers can see the uniqueness of Sabah's world—what she sees and what she feels in her daily life in Cairo, Egypt—but they can also relate to it because they are able to recognize in their own lives the same symbols, expressions, and interactions. I once asked Sabah why the majority of her photographic works seem to evoke sadness. She said that there are far fewer truly happy people in the world than those with problems. But at the same time, she said, the decoration and the rich colors evoke a cheerfulness that is in contrast to the sadness. Perhaps that is also why many people feel a connection with her work: it simultaneously expresses their struggles and their joys, or the pain and beauty of life.

Islam and the Image

You may be surprised by this story of a Muslim artist whose work focuses on people. Doesn't Islam ban all figurative images? Didn't the Taliban government of Afghanistan destroy the famous Buddha statues because they were an affront to Islam? Well, yes and no. Islam, like any world religion, takes different forms in different places, and many aspects of it are open to interpretation. The Taliban's interpretation of the place of the image in Islam is only one among many, and it is certainly not the most dominant. Sabah, like many Muslim visual artists around the world, believes that the injunction against image-making in Islam only applied to the early period of the religion, when there was a real danger that figurative images could be worshipped as idols. Concern about idolatry is found not only in Islam, but also in Judaism and some versions of Christianity. Over the centuries, Muslim artists from Morocco to Indonesia pioneered the use of floral and geometric designs that avoided the figurative image. They became famous for their innovations, and some Muslims say that their ancestors invented abstract art hundreds of years before Europeans and Americans. Today, however, images are pervasive in Muslim-majority societies all around the world, and are found in everything from art works to the mass media. It is the image in mass media that Sabah is interested in, and she is not opposed to it as a Muslim. Rather, she considers that it is partly her Muslim faith that leads her to want to respect and portray humanity in her art.

Art, Freedom, and Being a Muslim

You may also wonder how it is that a woman in a so-called traditional Muslim society was able to become a visual artist, speak at public events, be interviewed by the media, and travel around the world exhibiting her work. In the United States, people are often taught to believe that Muslim women, particularly those who wear a head covering, are more oppressed than non-Muslim women and have virtually no freedom. And art is often seen as the epitome of freedom. Certainly, Americans tend to think of artists as renegades, rebels, or outcasts—people who are not bound by social or religious mores. And since Muslim women are so often portrayed in the Western media as having no individuality, voice, or power, and as being bound to religious mores, the logical conclusion is that none of them could be artists. Yet there are thousands of female artists around the world who identify as Muslims. Many of them are in Egypt, where women have been doing fine art for nearly a century. Some of them wear their own different versions of the veil—anything from a long black chador to a short floral headscarf—but this piece of clothing is usually not the most important part of their lives. Some wear it because they believe it is a religious duty, some because they feel naked without it, some because their government mandates it, and some because it's fashionable. For Sabah, her veil is just one small part of the rest of her life, which is dedicated to God, art, and freedom.

In fact, she sees religion, art, and freedom as interrelated and as dependent on one another. To her, being an artist means living what she calls an "open, liberal life." This means being a good Muslim and not letting society's mores completely dictate her life. For example, women her age are usually married with children. They either don't work, or come home from work and spend their evenings with their families. Sabah has refused to live up to this expectation. She still lives at home with her family though, which is socially expected of unmarried women and is something that she does because she wants to. But unlike most other younger women who live at home, she stays out late at night, going to exhibition openings or talking with male colleagues at coffee shops. She also contributes significantly to the family's income by bringing home her money from art sales, and has started to be more of a breadwinner than her brothers (who would normally be expected to support the family as her father approaches retirement age). She says that leading this kind of lifestyle is necessary for her to be a freethinking artist who also cares about her family. And by doing so, she is following Islam.

Sabah explains that wearing the veil, praying, and fasting bring her closer to God. And she argues that "religion is ease, not difficulty." Many male artists in her cohort, along with most artists of the older generation, think that wearing the veil and being openly religious goes against the freedom necessary to be an artist. One male art professor once told me that he is disturbed by the rise in veiling among the younger generation of women in his classes. He said that he tells his students who wear the veil that art requires a person to be free, and that they can't have freedom if they cover their heads. Those artists who denigrate veiling, which is now practiced by the majority of urban Egyptian women, represent an older secular nationalist ideology that believed that women must discard the veil for Egypt to be considered truly modern. But many scholars have argued that this practice of veiling is actually very modern. Sabah's version of veiling emerged in the late 1970s when women started to enter the work force and higher education in greater numbers. It was a way for them to contribute to society, enter public life, develop themselves, and help support their families—without being compelled to compromise the value they placed on modesty and on retaining the importance of family support and protection in their lives. Many women in Egypt argue that the veil enables them to do things they wouldn't otherwise feel comfortable doing, such as working alongside unrelated men whose kinship ties may be unknown. Indeed, Sabah represents a new generation of women artists in Egypt who do not believe that one has to look more Western in order to be freethinking or modern. Sabah told me,

> I personally see no difference between a woman wearing a swimsuit and a woman wearing a veil, because what you wear is a personal thing. I decided to wear the veil three years before my mother did. So it was not forced on me. I even asked my father plenty of times about what he would think if I took it off, and he said to do it.

Her critics, like many non-Muslim Americans, think that it is Muslim men who force women to veil. Sabah's situation shows that the practice is much more complicated than that.

Sabah and others feel that the veil actually offers them more freedom from gender discrimination, because they are judged on the basis of their intellect rather than by standards of beauty. In response to her critics, she says,

> I feel that the veil is giving me so much freedom that they cannot see. It gives me complete freedom to deal with people through my intellect and not through a face or body. You could be beautiful and people treat you in a certain way as a result. I want people to treat me as a human being, and not as an object of beauty or ugliness.

See the Art, See the Person

Sabah got really mad at me one day. I had written an article about her for an academic journal in the United States. Before I sent the article off, I gave Sabah a copy for her comments. It was important to me, as a friend and as an anthropologist, to share my work with the person on whom it was based. A few days later, Sabah stopped by my apartment. It was clear that she was quite upset. She said that my article wrongly focused on the fact that she wears a veil, whereas it should have focused on her art. My heart started pounding, because she understood me to have done precisely what I didn't want to do. I had studied well the ways in which Middle Eastern women's head coverings had become fetishized by the French and British colonizers of Arab countries. For them, the veil was a symbol of the hypersexualized eroticism attributed to Middle Eastern women in their harems. Then and now, it was also used to blame Muslim men for oppressing women and thereby to justify military invasions of Muslim countries. This is part of the ongoing creation of an "Oriental Other" that Edward Said called Orientalism. I certainly did not want to be part of that tradition.

I also did not want to be like some of the other Westerners who were intrigued by Sabah because, in their ignorance of social life in the Middle East, they couldn't imagine how a veiled woman could do such great art. I also did not want to be like the curator from Europe who once asked Sabah to contribute her work to a show on the veil. Sabah asked her why she should be in such a show, since her work had nothing to do with the veil. It became clear that the curator wanted to include her only because of the fact that she wore a veil. She could not see Sabah as a person and as an artist with her own interests. Sabah and her art became reduced to a piece of cloth.

I told Sabah that I had tried not to focus on the veil, but that as an anthropologist I had to analyze social context as much as the art works. She had a copy of my article in her hand. I couldn't help but notice that she had gone through it with an English-Arabic dictionary to translate word for word, but she had only worked on the very beginning (where I mentioned the veil) and not the rest of it (where I discussed the art and why it's important). Clearly, language and general issues of translation, which occurred less with the more elite artists, were part of the misunderstanding. But whether or not I overemphasized the veil (I ended up changing the article a bit), the point is that Sabah gets very angry when she thinks that someone is paying more attention to the way she dresses than to her art, or gives less importance to her as a person rather than a stereotype of a "Muslim woman." She knows quite well the Western imagery of Middle Easterners that I mentioned earlier, and she wants no part of it. Indeed, one could argue that the current fascination with her veil is just an updated version of colonial-era Orientalism. Indeed, she feels as if she is in a Catch-22 situation, because while Westerners are intrigued by her veil, many older generation Egyptian artists and curators are repulsed by it. She even believes that she has experienced serious discrimination in the local art scene solely because people can't look past her veil.

The other problem I had while writing this article was that it was supposed to be for a journal on feminism, yet Sabah resists any feminist interpretation to her work. She is critical of gender discrimination in the home and workplace, but she is not concerned with that in her work—partly because she is aiming for a kind of universality. She told me that she is interested in portraying "the human state, regardless of gender." And when I asked her if her art contained feminist elements, she said defiantly, "I don't really feel that I am making women's art. I make art as art and it is not relevant to my gender." In part she was responding to the desire among many Western curators to see in Muslim women's art a critique of Islamic patriarchy. Today, most European and American exhibitions of art made by Muslim women get framed in this way. There is an assumption that Muslim women are inherently more oppressed than other women around the world, and that therefore their art should critique that oppression. Like Sabah, I think that women are oppressed in different ways everywhere, and that there is nothing inherent in any religion that makes women who practice it more oppressed than others. I also appreciate her persistent refusal to be classified as a Muslim woman artist who resists Islamic patriarchy. These kinds of classifications are more about the interests of Western art institutions, audiences, and curators than they are about the kind of art that Muslim women around the world do, or about the ways in which they practice and experience their religion. Furthermore, the desire to see Sabah's art, or the art of any other Muslim woman, as feminist resistance to Islam actually ends up silencing their voices and ignoring the diversity of their artistic expressions. So instead of forcing Sabah's art into this kind of framework, or chiding her for not being feminist enough, I tried to show why the work doesn't fit into such classifications and the reason for her stance.

When I was writing this chapter, I also struggled to find ways to describe Sabah as an Egyptian without reducing her complicated identity as solely attributable to that place. Whereas Muslim artists in the international, Western-dominated art scene face long-standing prejudices because of their religion, artists who live and work in non-Western countries such as Egypt are likely to be pigeonholed because of their national identity. While Sabah and her colleagues remain inspired by the material and visual culture of their country and are committed to it in a variety of ways, they do not like the fact that in the international arena they are always described first as "Egyptians" and then as "artists." (Sabah once asked why Picasso was never described as a "Western" or "Spanish" artist, while she always has to be described as "Egyptian.") She and her peers also often find themselves in the bind of having to do art that fits Western curators' idea of what is "Egyptian." At the same time, if their work looks "too Egyptian," then they are criticized for being "too provincial" or "too nationalist." Either way, they face ghettoization. Here, I have tried to show some of the ways in which her art belongs to a context that is Egyptian as well as international at the same time. Sabah Naim is a hometown artist of the world.

The problems of reception that Sabah has faced are gradually disappearing due to the strength of her work and the power of her voice. Hopefully, this is also a sign that the larger system of stereotypes and prejudices toward Egyptians, Arabs, and Muslim women are starting to break down. Americans often ask, "What will bring peace to the Middle East?" Sabah and her work suggest three answers to this banal question. First, Americans need to recognize the fact that they play an active role in reproducing visual and other stereotypes that feed the instability. Second, Americans would do well to empathize with their fellow human beings halfway around the globe, to see them as people who, despite being from individual worlds of their own, are part of the bigger world that we all share. And third, we need to avoid silencing the voices of Arabs and Muslims, and of women especially. We need to listen to what they have to say.

In that spirit, I'd like to let Sabah Naim end this chapter. In the spring of 2000, Sabah was about to embark on the first of what would be many exhibitions abroad in Europe. She also had a major show at a newly opened foreign-owned gallery in Cairo. She was excited, busy, and nervous about what lay ahead. I asked her how she hoped that she and her work would be received by her new audiences. She said,

> I hope that whoever sees my work feels it. If my work is able to reach their feelings, then I have succeeded in communicating my own. I want them to see me as me and that's it. Will they be able to accept me and understand me as an artist? Not as an Egyptian, or a veiled woman, or whatever, but as an *artist*....

Select Bibliography

Introduction

Ahmed, Leila. *Women and Gender in Islam: Historical Roots of a Modern Debate.* New Haven: Yale University Press, 1992.

Encyclopedia of Islam in the Muslim World, 2 vol. Richard C. Martin (ed.). New York: Macmillan Reference, 2004.

Winter, T.J. and John A. Williams. *Understanding Islam and the Muslims: The Muslim Family and Islam and World Peace.* Louisville, KA: Fons Vitae, 2002.

Grounded

1. Islam in the Balkans

Bringa, Tone. *Being Muslim the Bosnian Way: Identity and Community in a central Bosnian Village.* Princeton: Princeton University Press, 1995.

Donia, Robert J. and John V.A. Fine. *Bosnia and Hercegovina: A Tradition Betrayed.* New York: Columbia University Press, 1994.

Poulton, Hugh and Suha Taji-Farouki. *Muslim Identity and the Balkan State.* London: Hurst & Company, 1997.

Trix, Frances. "The Resurfacing of Islam in Albania." *The East European Quarterly* 28 (4) (1995): 533–549.

2. Islam in Morocco

Becker, Cynthia. *Amazigh Arts in Morocco: Women Shaping Berber Identity.* Austin, TX: University of Texas Press, 2006.

Fernea, Elizabeth. *In Search of Islamic Feminism: One Woman's Global Journey.* New York: Doubleday, 1998.

Geertz, Clifford. *Islam Observed: Religious Development in Morocco and Indonesia.* Chicago: University of Chicago Press, 1971.

Mernissi, Fatima. *Dreams of Trespass: Tales of a Harem Girlhood.* Reading, MA: Addison-Wesley Publishing Company, 1994.

Westermarck, Edward. *Ritual and Belief in Morocco.* New York: University Books, [1926] 1968.

3. Islam in Pakistan

Ahmed, Akbar S. Jinnah. *Pakistan and Islamic Identity: The Search for Saladin*. London: Routledge, 1997.

Buehler, Arthur F. *Sufi Heirs of the Prophet: The Indian Naqshbandiyya and the Rise of the Mediating Sufi Shaykh*. Columbia: University of South Carolina Press, 1998.

Hegland, Mary Elaine. "Shi'a Women's Rituals in Northwest Pakistan: The Shortcomings and Significance of Resistance." *Anthropological Quarterly* 76 (3) (Summer 2003): 411–442.

Jalal, Ayesha. *Self and Sovereignty: Individual and Community in South Asian Islam since 1850*. London: Routledge, 2001.

Nasr, Seyyed Vali Reza. *The Vanguard of the Islamic Revolution: The Jama'at-Islami of Pakistan*. London: I.B. Tauris Publishers, 1994.

Pinault, David. *The Shiites: Ritual and Popular Piety in a Muslim Community*. New York: St. Martin's Press, 1992.

4. Islam in Iran

Hegland, Mary Elaine. "Gender and Religion in the Middle East and South Asia: Women's Voices Rising." In *Social History of Women and Gender in the Modern Middle East*, edited by Judith Tucker and Marlee Meriwether, 177–212. Boulder, CO: Westview Press, 1999.

———. "Two Images of Husain: Accommodation and Revolution in an Iranian Village." In *Religion and Politics in Iran: Shi'ism from Quietism to Revolution*, edited by Nikki R. Keddied, 218–236. New Haven: Yale University Press, 1983.

Kamalkhani, Zahra. *Women's Islam: Religious Practice Among Women in Today's Iran*. London, New York: Kegan Paul International, 1998.

Loeffler, Reinhold. *Islam in Practice: Religious Beliefs in a Persian Village*. Albany, NY: State University of New York Press, 1988.

Moaveni, Azadeh. *Lipstick Jihad: A Memoir of Growing Up Iranian in America and American in Iran*. New York: Public Affairs (Persius Books Group), 2005.

Mottahedeh, Roy. *The Mantle of the Prophet: Religion and Politics in Iran*. Reprint edition. Oxford: One World, 2000.

In a New Place

5. Islam in Europe

Bowen, John R. *Why the French Don't Like Headscarves: Islam, the State and Public Space*. Princeton: Princeton University Press, 2006.

Ewing, Katherine Pratt. Stolen Honor: *Stigmatizing Muslim Men in Berlin*. Palo Alto: Stanford University Press, 2008.

Goody, Jack. *Islam in Europe*. Oxford: Polity (Blackwell), 2004.

Metcalf, Barbara Daly. *Making Muslim Space in North America and Europe*. Berkeley: University of California Press, 1996.

6. Islam in West Africa

Diouf, Mamadou and Mara A. Leichtman. *New Perspectives on Islam in Senegal: Conversion, Migration, Wealth, Power and Femininity*. Palgrave Macmillan, forthcoming.

Gellar, Sheldon. *Senegal: An African Nation between Islam and the West*. Boulder, CO: Westview Press, 1982.

Leichtman, Mara A. "The Legacy of Transnational Lives: Beyond the First Generation of Lebanese in Senegal." *Ethnic and Racial Studies* 28 (4) (2005): 663–686.

Momen, Moojan. *An Introduction to Shiʿi Islam: The History and Doctrines of Twelver Shiʿism*. New Haven, CT: Yale University Press, 1985.

Quinn, Charlotte A. and Frederick Quinn. *Pride, Faith and Fear: Islam in Sub-Saharan Africa*. Oxford: Oxford University Press, 2003.

Robinson, David. *Muslim Societies in African History*. Cambridge: Cambridge University Press, 2004.

7. Islam in Iraq

Chibli Mallat, *The Renewal of Islamic Law: Muhammad Baqer al-Sadr, Najaf and the Shiʿi International*. Cambridge Middle East Library, Cambridge: Cambridge University Press, 1993.

Nakash, Yitzhak. *The Shiʿis of Iraq*. Princeton: Princeton University Press, 1994.

Nasr, Vali. *The Shia Revival*. New York: Norton, 2006.

Walbridge, Linda S. (ed.). *The Most Learned of the Shiʿa: The Institution of the Marjaʿ Taqlid*. New York: Oxford University Press, 2001.

Reconciled

8. Islam in Afghanistan

Doubleday, Veronica. *Three Women of Herat: A Memoir of Life, Love and Friendship in Afghanistan*. London: Tauris Parke Paperbacks, 2006.

Hosseini, Khaled. *Kite Runner*. New York: Berkeley Books, 2004.

Lipson, Juliene G. and Patricia A. Omidian. "Afghans and Afghan Americans." In *Transcultural Nursing: Assessment and Intervention*, Fifth Edition, edited by Giger, Joyce Newman and Ruth Davidhizar. 394–409. Philadelphia: Elsevier Science, 2006.

Miller, Ken, Patricia Omidian, Abdul Samad Quraishy, Naseema Quraishy, Mohammed Nader Nasiry, Seema Nasiry et al. "The Afghan Symptom Checklist: A Field Study of Mental Health in Kabul, Afghanistan." *American Journal of Orthopsychiatry* 76 (4) (2005): 423–433.

Omidian, Patricia and Kenneth Miller. "Addressing the Psychosocial Needs of Women in Afghanistan." *Critical Half* 4 (1) (2006): 17–21.

Rashid, Ahmed. *Taliban: Militant Islam, Oil and Fundamentalism in Central Asia*. New Haven: Yale Note Bene, 2000.

9. Islam in America

Abdo, Geneive. *Mecca and Main Street: Muslims in America after 9/11*. New York: Oxford University Press, 2006.

Aswad, Barbara and Barbara Bilge (eds.) *Family and Gender among American Muslims*. Philadelphia: Temple University Press, 1996.

Gomez, Michael A. *Black Crescent: The Experience and Legacy of African Muslims in the Americas*. New York : Cambridge University Press, 2005.

Haddad, Yvonne and Jane Smith (eds.). *Muslim Communities in North America*. Albany: State University of New York Press, 1994.

Walbridge, Linda. *Without Forgetting the Imam: Lebanese Shiʿism in an American Community*. Detroit: Wayne State University Press, 1997.

10. Islam in Indonesia

Azra, Azyumardi. *The Origins of Islamic Reformism in Southeast Asia: Networks of Malay-Indonesia and Middle Eastern Ulama in the Seventeenth and Eighteenth Centuries*. Honolulu: University of Hawaii Press, 2004.

Feener, R. Michael. *Muslim Legal Thought in Modern Indonesia*. New York: Cambridge University Press, 2007.

Gade, Anna M. *Perfection Makes Practice: Learning, Emotion, and the Recited Qur'an in Indonesia*. Honolulu: University of Hawaii Press, 2004.

Ricklefs, M.C. *Mystic Synthesis in Java: A History of Islamization from the Fourteenth to the Early Nineteenth Centuries*. Norwalk, CT: East Bridge, 2006.

———. *Polarising Javanese Society: Islamic and other visions*, Singapore: National University of Singapore Press, 2007.

11. Islam in Egypt

Hirschkind, Charles. *The Ethical Soundscape: Cassette Sermons and Islamic Counterpublics*. New York: Columbia University Press, 2006.

Nasrallah, Yousry. *On Boys, Girls, and the Veil*. Seattle: Arab Film Distribution, 1995.

Starrett, Gregory. *Putting Islam to Work: Education, Politics, and Religious Transformation in Egypt*. Berkeley: University of California Press, 1998.

Winegar, Jessica. *Creative Reckonings: The Politics of Art and Culture in Contemporary Egypt*. Stanford: Stanford University Press, 2006.

Notes on Contributors

Cynthia Becker has lived and studied in Muslim Africa, including Morocco, Algeria, Tunisia, Senegal, Niger, and Mali. Her book, *Amazigh Arts in Morocco: Women Shaping Berber Identity* (2006), considers the evolution of Amazigh (Berber) arts in southeastern Morocco from the early twentieth century to the present and women's roles in artistic production. She is currently an assistant professor of art history at Boston University.

Katherine Pratt Ewing has lived and done research in Pakistan, Turkey, and Germany. Her book *Arguing Sainthood: Modernity, Psychoanalysis and Islam* (1997) is based on nearly two years of anthropological fieldwork among Sufi saints and their followers in Pakistan. She is involved in ongoing research among Turkish Muslims in Germany and among South Asian Muslims in the United States. Dr. Ewing is currently associate professor of cultural anthropology at Duke University in Durham, North Carolina.

R. Michael Feener has lived, worked, and studied in Indonesia, Egypt, and the Yemen and made shorter visits to Muslim communities in other parts of the Middle East and Africa. His books include *Islam in World Cultures: Comparative Perspectives* (2004), *Muslim Legal Thought in Modern Indonesia* (2007), and *Islamic Law in Contemporary Indonesia: Ideas and Institutions* (2007). He is currently associate professor of history at the National University of Singapore and a senior research fellow at the Asia Research Institute.

Mary Elaine Hegland has conducted fieldwork in Iran, Pakistan, Tajikistan, Turkey, and Afghanistan, and among Iranians in northern California. She has published on religion, ritual, and gender in an Iranian village during the Revolution of 1978/1979, and more recently on aging and the elderly in Iran and among Iranians elsewhere. She coedited *The Islamic Resurgence in Comparative Perspective* (1987) and a special issue on ethnography for the *Journal* *of Iranian Studies* (2004). Dr. Hegland is associate professor of social-cultural anthropology at Santa Clara University in California.

Mara A. Leichtman has lived in Israel, Morocco, Lebanon, and Senegal, and has worked with Muslim communities in metropolitan Detroit, Michigan, and in London, England. Her dissertation is entitled "A Tale of Two Shi'isms: Lebanese Migrants and Senegalese Converts in Dakar." She is assistant professor of anthropology and Muslim studies at Michigan State University.

Patricia A. Omidian has lived and worked among Muslims in Iran, Pakistan, and Afghanistan, as well as among immigrant Muslim communities in the San Francisco Bay area. She wrote *Aging and Family in an Afghan Refugee Community* (1996) on the Afghan community in California. She also worked with the Palestinian and Iranian communities in San Francisco. Dr. Omidian was the Country Representative for Quaker Service Afghanistan (American Friends Service Committee) and resided in Kabul, Afghanistan for five years. She is currently based in Pakistan.

Frances Trix has lived and worked among Muslims in Turkey, south Lebanon, Yemen, Kosova, and Albania, as well as among immigrant Muslim communities in metropolitan Detroit. She wrote *Spiritual Discourse: Learning with an Islamic Master* (1993) from her ongoing study in a Bektashi Muslim community. Dr. Trix is currently associate professor of linguistics and anthropology at Indiana University in Bloomington, Indiana.

John Walbridge has lived in Lebanon, Jordan, Pakistan, and Turkey. He is the author of several books on the Illuminationist school in Islamic philosophy. He also works on philosophical aspects of Islamic medicine. Dr. Walbridge is presently professor of Near Eastern languages and cultures at Indiana University, where he was recently chair of the department.

Linda Walbridge was a cultural anthropologist specializing in religion and had worked in Lebanon, Jordan, Indonesia, and Pakistan. She wrote books on the religious life of the Lebanese Shi'a living in the Detroit area and on the Christians of Pakistan. She also edited a volume on alternative Shi'ite leadership. She died in 2002.

Jessica Winegar has spent nearly four years living in Egypt and doing research with Muslim artists and intellectuals. She has written numerous articles on art and visual culture in the Middle East and is the author of *Creative Reckonings: The Politics of Art and Culture in Contemporary Egypt* (2006). She also works with museums and cultural organizations seeking to promote Arab and Muslim artists in the United States. Dr. Winegar is currently an assistant professor of anthropology at Temple University.

Index